The Electric Guitar

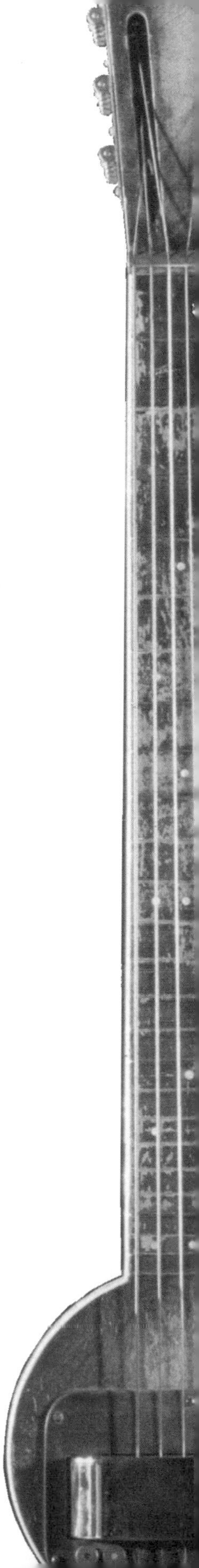

The Electric Guitar

a history of an american icon

edited by André Millard

For The Lemelson Center
for the Study of Invention
and Innovation
National Museum
of American History
Smithsonian Institution
Washington, D.C.

The Johns Hopkins University Press
Baltimore and London

Printed in the United States of America on acid-free paper
9 8 7 6 5 4 3 2 1

The Johns Hopkins University Press
2715 North Charles Street
Baltimore, Maryland 21218-4363
www.press.jhu.edu

Library of Congress Cataloging-in-Publication Data

The electric guitar : a history of an American icon / edited by André Millard.
p. cm.
"The book is drawn from a 1996 symposium on the invention of the electric guitar presented by the Lemelson Center for the Study of Invention and Innovation, National Museum of American History, Smithsonian Institution"—ECIP data.
Includes bibliographical references (p.) and index.
ISBN 0-8018-7862-4 (hardcover : alk. paper)
1. Electric guitar—History—Congresses. I. Millard, A. J.
ML1015.G9E43 2004
787.87'1973—dc22 2003016415

A catalog record for this book is available from the British Library.

Photographs for pages i, iii, and xi, and at all chapter openers, courtesy Smithsonian Institution.

contents

preface

My first and greatest debt of gratitude is to the contributors, whose splendid work you will soon be reading. Charlie McGovern, Susan Schmidt-Horning, James P. Kraft, and Rebecca McSwain were all participants in the Smithsonian's "Electrified, Amplified, and Deified" symposium about the electric guitar—the starting point of this book. John Strohm, both historian and indie legend, joined the project later. Arthur Molella and his staff at the Lemelson Center for the Study of Invention and Innovation are to be congratulated for putting on such an ambitious and successful symposium and for supporting the development of a volume based on its exploration of the meanings of the electric guitar. Alison Oswald provided archival assistance, Joyce Bedi and Monica Smith worked with the photographs, and the whole project was expertly administered, through thick and thin, by Claudine Klose.

The process of reviewing and editing this manuscript took several years. Over time it became evident that we did not have enough material in the original papers given at the Smithsonian symposium to make up an acceptable history of the electric guitar—there was no mention of the Fender Stratocaster for example. Similarly the important topic of amplifiers and effects was raised in the symposium but hardly mentioned in the papers. As editor I began to fill in the gaps, and two short transitional pieces that were intended as a link between papers became chapters 4 and 6. Rebecca McSwain's original slide-show presentation at the Smithsonian was not deemed appropriate for publication, so after the symposium we worked together on a paper about the guitar hero. During the last three years I turned this collaborative effort into chapters 7 and 8 as I drew upon the input of several other scholars and musicians. John Strohm was recruited to cover the years of alternative rock in chapter 9 and bring a musician's viewpoint to the material. When the person who agreed to write chapter 10 dropped out at the last minute, I reluctantly filled in. The long development of this manuscript after the symposium was mainly in my hands and I take responsibility for any errors that might have crept into the text.

Many musicians, inventors, and businessmen agreed to be interviewed as part of this project. I want to thank them for their time and expertise. Robert J. Brugger of the Johns Hopkins University Press was more than an editor—he was also a facilitator, organizer, and booster. Thanks also to Melody Herr, who shepherded this volume through the many years of its development. I owe a great deal to Harris Berger, who contributed his invaluable musical knowledge, sharp eye, and incisive intellect as I worked through the drafts. My guitar-playing buddies Bryan Price, Eric Erbskorn, and Scott Smith gave me the benefit of their extensive knowledge. I want to thank my friends Mike Wood and Ted Scannell for their information and inspiration about the English scene.

At my home institution of the University of Alabama at Birmingham I owe thanks to reference librarian Brian Brookshaw, my colleague Andrew Keitt, my graduate students Kurt Kinbacher and Rob Heinrich, and, most important, that Escalade-driving Top Dog Debbie Givens.

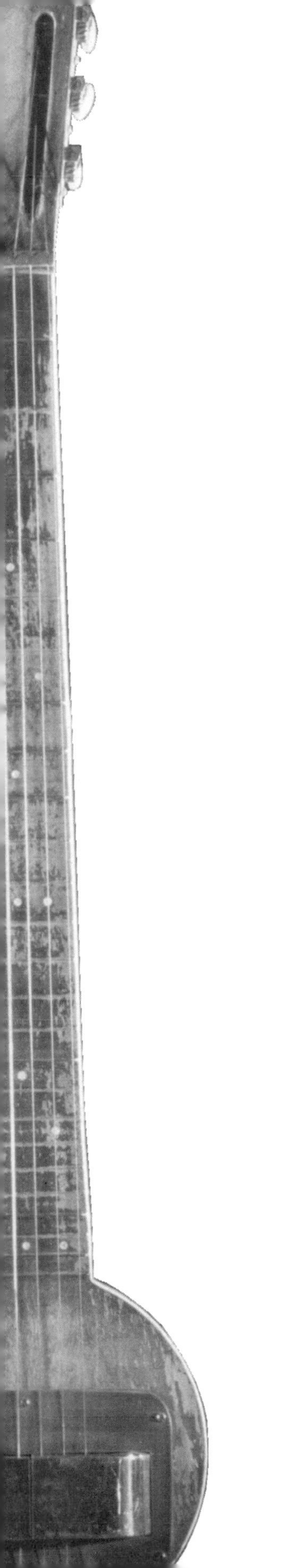

The Electric Guitar

introduction

american icon

On the cover of a book of readings about the Americanization of immigrants, Philip DeVita and James D. Armstrong's *Distant Mirrors: America as a Foreign Culture* (1988), a collage of various images suggested the essential United States and its lifestyle. They included a cheeseburger, several skyscrapers, a yellow taxicab, the Statue of Liberty, and an electric guitar—not the acoustic instrument, but a shiny, colorful solidbody electric. An edition of 2002 shows only a flag, a basketball player, a skateboarder, a skyscraper, and the guitar. By this measure the electric guitar has been added to the symbols of the United States, along with such easily recognized items as the American eagle, the Cadillac, and the diner.

There must be something about this instrument that strikes foreigners as American: its ready availability as a item of consumption in an affluent society, its central role as an entertainer in the youth culture, and its high degree of technological sophistication in an advanced industrial state. It is one of those cherished machine-made possessions, like automobiles, that Americans use to define themselves. Alexis de Tocqueville, the first great observer of the American way of life, found a pragmatic, materialistic, restless nation fixed on the pursuit of wealth and firmly attached to the idea of technological progress. The Americans he described in the early nineteenth century had little use for the aristocratic high culture of Europe; they preferred to entertain themselves with more democratic pursuits. This volume will show how well the electric guitar and the mythology of rock 'n' roll fitted these perceived American characteristics. For these and probably many other reasons, retailing giants like Kmart, Burger King, Target, and Office Depot use twangy guitar sounds for their television advertisements. The Applebee's restaurant chain ends its guitar-laden TV spot with the thought that its food is "as American as apple pie." So, it seems, is the electric guitar.

As a signifier of Americanization it can even identify specific locations within the United States. The hard rocker Alice Cooper is proud to be from the city of Detroit, saying, "I'll never be in a band unless it has at least two guitars. I don't go for the keyboards/horns thing. I want big loud guitars. That's pure Detroit."[1] A television advertisement for the Honda Acura identifies a location as New Orleans and then plays a hazy, echoing guitar sound to reinforce that idea. The Honda Accord's pitchman is an elderly African American blues player who talks to the viewer as he plays his guitar in a bluesy, down-home style. The distinctive chugging rhythms of the blues, played with a slide guitar, never fail to conjure up images of the Deep South.

The sound of the electric guitar as a signifier of America was only one of the meanings of this instrument discussed at a symposium held at the Smithsonian Institution in 1996. Entitled "Electrified, Amplified, and Deified: The Electric Guitar, Its Makers, and Its Players," it brought together musicians, engineers, businessmen, and academics to investigate all the different uses and meanings of the electric guitar. The symposium encompassed scholarly presentations, roundtable discussions with people who made significant contributions to the development of the technology, and performances by renowned guitarists. All these different points of view produced a multifaceted portrait of the guitar that went well beyond its role as a musical instrument.

The papers presented at this program addressed issues as diverse as American technological enthusiasm, the differences between craft and mass production techniques, and the guitar as a symbol of masculinity. This book brings together some of the papers given, although it cannot do justice to the varied group who participated in the symposium. Any project supported by the Smithsonian and the Jerome and Dorothy Lemelson Center for the Study of Invention and Innovation must see the guitar primarily as an artifact of the inventive process. Consequently, this volume dwells on the technology of electrification and considers the guitar and its amplifier as a technological system. The contributors show that innovation occurs in varied contexts and is carried out by many different people. The history of the electric guitar demonstrates a wide variety of innovative behavior, from the tinkerers who devised the first electric pickups to the studio session players and engineers who rigged up the unique psychedelic sound of the 1960s.

Today's familiar, man-made sounds run from the roar of an internal combustion engine to the "canned music" coming from clever but imperfect copies of long-lost performances. The electric guitar generates one of these sounds. It emerges from countless entertainment machines that occupy our homes, our cars, our public spaces, and our workplaces. Everyone recognizes its distinctive amplified sound. It is hard to avoid, because it serves the entertainment business in many ways: as the signature of popular music, in soundtracks for motion pictures, and as music for advertising on radio, television, and the movies. It has become one of the unavoidable sounds of modern life—as common in our global popular culture as the ring of a telephone.

The electric guitar is a versatile instrument capable of many different sounds, and some of them are so well known that they achieve instant recognition. For example the distinctive glistening metallic notes of surf music (a style of guitar music made popular in the 1960s) pop up in environments as diverse as advertisements for fast food restaurants, television shows (*Third Rock from the Sun*), and the movies (*Pulp Fiction* and *Twelve Monkeys*). Many of today's younger listeners might not even know of its origins in garage bands way back in the early 1960s, but they would certainly recognize this sound and assign it meanings different from those created over forty years ago.

american business and invention

The ubiquitous guitar sound in our culture is partly the result of the guitar's success as a consumer product. About 700,000 electric guitars are sold every year. Sales took off in the early 1960s, when rock 'n' roll was in its heyday and the number of guitars sold each year jumped from 300,000 to around 600,000. The number of guitar players in the United States has increased steadily since the 1960s: an estimated 7 million in the 1970s, 9 million in the 1980s, and over 10 million predicted for the first decade of this century.

Manufacturing electric guitars is part of the very big business of entertaining Americans. It began in earnest after World War II and grew rapidly in the 1950s and 1960s. It is a high-tech industry driven by constant innovation and sometimes disrupted by unexpected changes caused by new technology or changes in fashion. The manufacturers have mastered the techniques of mass production to bring us affordable, sophisticated guitars in every shape and color. We now have hundreds of models to choose from, and at a wide range of prices. Most of the big American guitar companies offer a complete package of guitar, amplifier, tuner, power cord, plectrum, and music lessons in one box for less than $500. Some foreign manufacturers offer the same type of starter kit for about $300. It is possible to buy an electric guitar for less than $100, especially at the musical instrument superstores that offer hundreds of guitars at discounted prices.

The business of selling guitars has been subject to the same changes that have decimated America's main streets and moved retailing to superstores in malls. The friendly local music stores—either delightful or intimidating depending on how well one can play the instrument—are a dying breed, along with the pony-tailed ex-musicians who usually man the counters. Acquiring an electric guitar is now as easy as buying an automobile battery or a vacuum cleaner. Musical instrument superstores that emulate the marketing strategies of Super-K or Wal-Mart display hundreds of cheap guitars in one large space. For those unwilling to drive to the mall there are several catalog companies (often linked to the chains of guitar superstores) that will fill phone, postal, or electronic orders.

There is an instrument available for every kind of customer. For the very young there are unbreakable plastic toy guitars. Older children start learning to play on scaled-down models, some of which come with a small amplifier. Some have the amplifier circuits built into the body. For the footloose there are lightweight, miniature travel versions. The shape of the electric guitar is also marketed in watches, candies, ornaments, neckties, T-shirts, and calendars. You can buy a guitar clock to put on the wall with your guitar sculpture and your tribute to Elvis—sacred pictures and text arranged into that familiar shape.

The development of the electric guitar is an excellent example of the ways technological innovations are applied to everyday products. Its sleek body incorporates the latest in electronics, metallic paints, and synthetic materials. Its invention is a part of the technological advance that created a new type of industrial society in the United States. Its power comes from electricity, a mysterious force introduced to Americans by their hero Thomas Edison in the nineteenth century. Edison's customers both welcomed and feared the new technology, believing that it could stimulate mental power, psychological energy, and sexual attraction but mindful that it was extremely dangerous. It could be gently applied to aches and muscle strains, but a great surge of current could energize Frankenstein's monster, much as electric guitars energized the stale popular music of the 1950s. Just like electricity, the electric guitar is so versatile and ubiquitous that it can have conflicting meanings.

This book examines some of its applications and meanings. Like all technology, the electric guitar comes with advantages and disadvantages. On one hand it is the most democratic of instruments, available to all and not too expensive. On the other hand the power of the amplified guitar is so overwhelming that it drowns out many of the other instruments that used to be commonplace in our lives. When was the last time you heard the sound of an accordion or a harmonica? The electric guitar is as much a symbol of noise pollution as it is of youth and excitement. One pharmaceutical company uses it to sell headache medicine, reminding us that there is "a fine line between music and noise."

As an example of the wonders of new technology, the electric guitar is a symbol of progress. Its shape and electrical components evoke modernity and high technology, but its sound, its ability to produce a level of volume unknown in the natural world, defines it as a symbol of humankind's technological dominance over the environment. Bluesman Brownie McGhee once said that he played an electric guitar because he believed in progress. His contemporary the late John Lee Hooker also loved electricity: "You barely have to touch the guitar, and the sound comes out so silky. Electric sound is so lovely."[2] Not everybody loved this sound, and in some contexts it appeared as an intrusive, disruptive force. When folksinger Bob Dylan replaced his acoustic guitar with an electric at the 1965 Newport Folk Festival, his audience was outraged. They accused him of selling out his folk roots and betraying both his music and his fans. A music once shared with the audience had been transformed into an overbearing commercial sound that drowned out individual voices—all the result of plugging in!

american music

Musicians were amplifying guitar sounds as early as the 1930s, but it was not until the 1950s that the electric guitar made itself heard. This was a consequence of the rise of rock 'n' roll. Perhaps because of the social impact of this new popular music, and the great changes occurring in American society at that time, the significant developments of the 1930s and 1940s have been overlooked—submerged in the critical and academic attention given to rock 'n' roll and the rise of a youth culture. Electric guitars, especially the solidbody models, played a central role in the story of rock 'n' roll and its unassailable dominance of popular entertainment. Their flexibility as musical instruments, combined with low price and easy portability, marked a new era in the business of entertainment as well as its sound.

This mass-produced consumer good represents the benefits of an industrialized society that transforms rare and expensive musical instruments into affordable music makers. From the 1950s the electric guitar has been a mainstay of the entertainment industry, not only by equipping every rock band that has come along but also by appearing in countless bedroom and garage practice spaces. One of the ways technology has transformed our lives has been by its everyday presence in them, bringing an unending stream of novelty and wonder. The two-hundred-dollar electric guitar is an example of this process in that it has brought a technology once limited to professional musicians into the reach of anybody, even a person with little musical training or talent. Writing about the impact of the phonograph in the nineteenth century, the historian Daniel Boorstin argued that it was "as good a symbol as any of the particular intrusion of technology in America into the traditional categories of culture." The electric guitar achieved this status in the twentieth century because, like the phonograph, it was a "democratic instrument."[3] Unlike its great competitor the piano, it was cheap to acquire, easy to learn, and simple to play. And this might explain why it went on to dominate the sound of American popular music.

Learning to play an electric guitar has become a rite of passage for thousands of teenage boys. Playing in a successful rock band was one of those universal dreams of the baby boom generation. It was part of that dance that we all did in preparation for an imagined life in the spotlight of popular entertainment. This hope, based on the rags-to-riches mythology of the American dream, gave rock 'n' roll its particular power as a cultural force. It embraced its audience with a comforting wall of sound and gave each person the germ of hope of transforming his or her life from the ordinary to the extraordinary. Of course very few made it to the top of the entertainment industry, but that fact never stopped anyone from dreaming about it. Possessing a guitar is part of this fantasy, even if you can hardly play it. The electric guitar was as much a part of the imagination of the 1950s and 1960s as it was part of the music.

The sound of the electric guitar has an immediate and persuasive impact on young men. Consider this picture painted by Roddy Doyle in his novel *The Commitments,* in which a bunch of young Dubliners experience the redemptive power of rock 'n' roll. One of them has just put on a record by a band called the Byrds: "Thirty seconds into the song the lads wanted to be the Byrds. They'd been demolished by the rip roaring guitars. . . . It was sweet and rough at the same time. . . . They didn't just hear it either. They were in its way. It went through them. Man's music."[4]

The amplified guitar sound—the distinctive crash of the power chord or the screeching whine of distortion—is the signature of rock 'n' roll. Once recognized, it immediately tells the listener what to expect next; fifty years of innovation and experimentation has exhausted nearly all the surprises in this genre. Whatever the current fashion in pop music, and whatever new electronic devices attract the attention of musicians, the electric guitar still rules.

Yet the sound of the rock guitar is rarely the unadorned output of the instrument alone but rather a processed sound that depends on the amplifier and includes the manipulation of numerous electronic effects. The musicologist Robert Walser once pointed out that the electric guitar was an incomplete instrument, something that had to be developed by musicians and technicians. In a small way this volume attempts to chronicle this completion by looking at technological developments in amplifiers, recording studios, and effects boxes and by acknowledging some important innovators who were not musicians. The engineer who thought out a way to incorporate echo into the sound of the guitar was as important as the musician who actually played it.

Most of the guitar sounds in our world are recorded; what we normally hear (unless we attend a live concert) is a recording of a guitar playing, and this technological wonder is the product of a long and complex system. The recording studio and the radio station have been the vital intermediaries in this technological network, and their effectiveness as diffusers of culture has ensured the dominance of the electric guitar in our popular music. The emergence of rock 'n' roll coincided with some important technological and economic changes in sound recording, all of which played a critical role in its creation and diffusion.

The term *rock 'n' roll* not only signifies a new type of popular music after World War II but also acts as a shorthand term for some significant social changes. George Brown Tindall's well-respected textbook, *America: A Narrative History* (1984–99), gives it a prominent place in the turbulence of postwar society and assigns it a major role in the youth rebellion of the 1950s and 1960s. Rock culture embraced new dances, new fashions, new methods of self-expression, and a new aesthetic. The electric guitar would be an important cultural artifact purely for what it says about our love affair with affordable technology and loud dance music. Many books have been written about its history and its profound affect on popular music. Yet there is more to the electric guitar than its sound; it has become one of the prime images of youth culture, occupying a critical space in the mythologies of popular entertainment and our hallowed tradition of rebellion.

A man playing the electric guitar appears and reappears on the packaging of recordings, in motion pictures, in music videos, and in millions of photographs of live performances. This image was soon co-opted by advertisers in the 1960s who were looking for a way to harness the energies of the counterculture and to commodify "hip." The so-called revolution in music could easily stand for all the other revolutions of the 1960s, and as a image of rebellion the leather-jacketed, guitar-playing rock star was hard to beat. In this way the electric guitar became a symbol of rebellious youth culture, and its player was depicted as a turbulent individual defying the stale status quo of the consumer society. As Thomas Frank points out in his book *The Conquest of Cool,* this image became one of the paramount clichés of our popular culture. Our televised marketplace is "a showcase of transgression and inversion of values, of humiliated patriarchs and shocked puritans, of screaming guitars and concupiscent youth."[5]

The shape of the guitar alone may serve as a symbol of popular music since 1945—such is its prominence in our visual landscape. The Rock and Roll Hall of Fame and Museum is the central point of homage to this cultural movement in the United States. It is housed in a magnificent building in Cleveland, Ohio, overlooking Lake Erie. The work of the renowned architect I. M. Pei, its design evokes the fingerboard of a guitar in the glass-and-steel-framed wall that leans up against the two main towers. Whether it is embodied in a skyscraper or in an MTV video, or even in the grillwork of the gate at Elvis's Graceland mansion, the image of the electric guitar has significant symbolic power.

american icon

The electric guitar has often been described as an icon of popular culture in the twentieth century, and certainly when newspaper editors thought about appropriate images to print at the turn of the millennium, the electric guitar came immediately to mind. Thus the *Manchester Guardian* promoted its end-of-century roundup of popular culture with a picture of Elvis Presley framed by Scotty Moore playing his Gibson ES-295 guitar.

For an icon to be worth its salt it has to have a powerful recognition factor. It has to be immediately recognized all over the world, like Mickey Mouse or Elvis Presley, for example. Most important, it has to say something, define some values or provide meaning. The electric guitar is an icon not only because it is all around us but also because it can stand for things other than musical sounds. We have already recognized that it can stand for technological progress and is the dominant image of rock 'n' roll. But what values can the guitar signify? Primary are those of the youth culture that once embraced rock 'n' roll. And if we want to investigate these values, one place to find them is on television, not in videos on MTV, but in the soundtracks of the advertisements on the broadcast networks.

Advertisers jumped on the electric guitar bandwagon in the 1960s as the means to evoke the excitement of youth and its intoxicating power. Its amplified

Guitars in our visual culture! This billboard is in downtown Atlanta, but it could be anywhere. Photo by M. Bodapati.

sound is still perceived as the language of transgression and hipness. Thus a basic distorted guitar riff, such as the introduction of "Rock'n'Roll Part 2" (recorded by Gary Glitter and now used to sell Mercedes cars), gives a sense of power, rebellion, and exhilaration. The Mitsubishi automobile company uses Curtis Mayfield's funky guitar introduction to the movie *Superfly* to sell its products. This sound is strongly embedded in the 1970s "blaxploitation" urban black scene and brings with it a heightened sense of hipness and counterculture. Not all electric guitar sounds need to be edgy and dangerous to sell products. The blue notes associated with B. B. King, which persuade us to eat Little Debbie cakes, are friendly, familiar, and relaxing. They suggest a rural rather than a tense urban environment and give a sense of well-being rather than drama.

Although advertisers as diverse as financial services and snack cakes employ the electric guitar sound to sell their products, the makers of automobiles have a special relationship with it. The automobile is a bastion of American consumer culture and a repository of some distinctly American values. Some rock songs come with a built-in message as well as appropriate guitar sounds: "Free Ride" (used to promote Buick automobiles) or Queen's rock anthem "We Are the Champions" (Ford trucks) speak for themselves. Bob Seger's "Like a Rock" not only makes a positive statement about the durability of Chevrolet trucks but also exploits Seger's rock credentials as a tough, traditionalist, blue-collar kind of a guy (from Detroit)—the person who might be interested in buying a truck, or the person the owner of such a truck might want to be associated with.

The sense of power implicit in the volume and attack of the electric guitar has served automobile advertisers very well. Certain genres of music are associated with driving, and loud distorted guitar sounds imply power and freedom. The drama of the Who's booming power chords gives the most sedate sedan an air of excitement and even menace. When General Motors cast around for suitable music for advertisements for its monstrous Cadillac SUV the Escalade, "the most powerful SUV on the planet," it unsurprisingly came up with the hard rock sound of Led Zeppelin.

Sellers of automobiles often focus on the journey to imply that their products bring freedom to drivers, employing guitar-driven songs like "Riding Down the Highway" (Doobie Brothers), Steppenwolf's "Magic Carpet Ride," and "Roller Coaster" (Rick James). Freedom in general has a powerful attraction for advertisers: "I'm Free" (by the Who and the Soup Dragons) is a very popular soundtrack in car advertisements. "Set Me Free" (the Kinks) has been used to sell Hundai automobiles. Consumer products like SUVs, personal computers, and stereo equipment are also depicted as "freedom machines" in advertisements, and the electric guitar helps make the point.

Since the 1960s advertisers have used symbols of youth culture and the sound of youth music to attract young customers or to imply novelty and hipness in products aimed at adults. In 2002 Nissan and Mitsubishi used fashionable grunge-inflected guitar sounds in their advertisements aimed at the twenty-something demographic, while staid American manufacturers like General Motors plundered the Top 40 to make Pontiacs and Buicks more appealing. Nevertheless, 1960s and 1970s rock music remains very popular, because baby boomers are still a very important group of consumers who are especially receptive to the appeal of guitar rock. There is nothing like the wail of a loud distorted guitar to encourage consumers of a certain age to reach for their credit cards. A television advertisement for the First Tennessee Bank shows a middle-aged baby boomer eyeing a shiny new motorcycle in a store. The guitar chords swell and reach a crescendo: Buy it! Buy it!

Some advertisers appropriate the lyrics or sound of a song without much consideration of its cultural context. Thus the Surfaris' classic "Wipeout," which once referred to slackers in 1960s surf culture, now helps sell Mr. Clean products. Ironically, Sly and the Family Stone's paean to excessive drug use, "I Want to Take You Higher," has been adopted by safe, sedate AT&T to impress its customers with the level of its service. The crunching guitar sound of the Troggs's "Wild Thing," an early excursion into heavy metal and an invitation to do really bad things, now lures older Americans to go on cruises with its siren sounds. The U.S. Postal Service uses Steve Miller's "Fly like an Eagle" to try to persuade us that it can deliver the mail.

Peter Townshend of the Who has acknowledged the "inestimable wealth" gained by "selling my music to Japanese car companies," aware perhaps of the possibility of devaluing the band's reputation by allowing advertisers to use its music.[6] But rock 'n' roll has sold out, and nothing illustrates the commodification of youth rebellion better than the continued exploitation of the electric guitar sound to sell goods and services. Ironically, and perhaps predictably, the sound and symbol of youthful rebellion has been taken over by the same capitalist forces that the musicians once hoped to undermine.

Playing guitars was once a solid masculine preserve, an activity for men. In a series of advertisements sponsored by the Seagram whiskey company entitled "It's what a man does . . . ," "at least one attempt to learn to play a guitar" was one mark of manhood (December 1998). Yet American society has changed drastically since the 1960s, and now playing an electric guitar is as much an activity of girls as of boys. "You've come a long way baby" was once applied to women who smoked cigarettes, but now it is applicable to how far female guitar players have come since the inception of rock 'n' roll. In an advertisement for Ovation Guitars, which promotes the message "plug into life," a young woman is seen playing the guitar. The venerable Martin company, one of the oldest guitar manufacturers in America, now uses a woman in its advertisements, and not a folksinger but a young professional seeking "creative balance in her high-pressured new lawyer's life."

Plugging into life has been an important part of teenage culture since the 1960s, when the equipment was made available to amateur players. The electric guitar has become an object of desire, something to yearn for, a prized possession with totemic powers. The movie *Wayne's World* celebrated the life of two "dweebs" who run a cable-access television show from their home. An electric guitar is the constant prop for Wayne, and his dream is to possess a 1964 Fender

Stratocaster (his mantra is "It will be mine"), which sits in a glass case in a music store. The white guitar plays a kind of Excalibur role in the plot—a sacred, symbolic object that empowers the user; only this is not the Arthurian legend but daily life in an American city. When the camera moves in on the guitar, sacred music plays. Eventually Wayne sells his soul (and his cable show) for the guitar. The guitar as a consumer dream is not confined to American life in the 1990s. The mean streets of Kingston, Jamaica, form the locale for *The Harder They Come,* the story of the rise to stardom of a reggae musician. The hero is a man with a mission to succeed, and we often see him looking into a shop window at a new, American-made electric guitar, which sums up all his hopes and dreams of a career in popular music.

Even if you never had a hope of learning to play and beginning that ascent to stardom, there was still a way to plug into life, an alternative in the form of the air guitar—that invisible guitar that anybody can play well, anybody who has memorized the moves and knows how to pose with the imaginary instrument. It is featured prominently in the movie *Risky Business* when the young Tom Cruise—at that time the everyman teenager—slides down the stair rail and starts playing air guitar to the background roar of Bob Seger's "Old Time Rock & Roll." His parents have left for a vacation and now he is in control of his world. The electric guitar is often depicted in movies as the great empowerer of American youth.

The great mass of players and listeners of the baby boom generation have carried their identification with the electric guitar into middle age. Most of them grew out of the rock 'n' roll dream and made a success of their lives in other endeavors. For example a guitar player in a band once called Ugly Rumours is now Prime Minister Tony Blair of the United Kingdom. Guitar music has a special appeal to this generation, and that is why it is so popular with advertisers. Forty years down the road it has lost none of its evocative powers. In the movie *American Beauty* a downtrodden middle-aged suburban husband rebels against his fate. The newly liberated Lester Burnham, in his moment of triumph, plays air guitar in his car to the sound of the Guess Who's "American Woman."

american art

Stripped of all its cultural meanings, the electric guitar is still a beautiful object in itself. Some people collect guitars just for the sake of enjoying their look and appreciating their rarity and value. Many of the guitars handmade at the Fender Custom Shop are destined for museums or glass display cases in private homes.

Gibson's Custom, Art, and Historic Division has produced a limited-edition Playboy Les Paul guitar, retailing for five thousand dollars. Only fifteen have been made. Perhaps Hef will buy one for the Playboy mansion. As an exclusive work of art it will probably not be played by a working musician.

Prized by collectors and valued by players, antique guitars—those dating from the 1950s and 1960s—now fetch premium prices. The great auction houses like Christie's and Sotheby's regularly hold auctions of rock 'n' roll memorabilia in which guitars are the most expensive and most desirable items. A guitar used by Scotty Moore on the famous Elvis records can be sure of a price tag exceeding one hundred thousand dollars. The value of a historic guitar used by a virtuoso player like Eric Clapton or Jimi Hendrix can be as much as a half-million dollars. Eric Clapton's 1956 sunburst Fender Stratocaster, called "Brownie," the one he used to record "Layla," is one of the most expensive guitars ever sold. There is a large and active market for antique instruments in the United States, and buyers come from all over the world to acquire a vintage Les Paul or Stratocaster.

As an objet d'art that exists as a beautiful possession, an ornament, the electric guitar can also be regarded as an example of the finest craftsmanship that displays rare skills and materials. The Chinery Collection sponsored a series of blue archtop guitars made by some of the outstanding guitar makers. The only stipulation was the kind of blue pigment used, and even so none of the guitars looked alike. The Smithsonian Institution's National Museum of American History displayed some of these guitars in a very popular exhibition in 1997–98 (see plate 1).

The word *classic* used to be reserved for the Romans and the Greeks and the orchestral music of high culture, but now it has been taken over by rock 'n' roll; we have classic rock radio stations playing classic tunes by classic bands who, naturally, play classic guitars. In an advertisement paid for by BMW (the German automobile manufacturer) a picture of a 1956 Gibson Les Paul electric guitar was described as "one rock classic that sounds better everyday." The point of the advertisement was that few products improve over time and the electric guitar (and presumably the right automobile) is one of them. The makers of electric guitars are part of an elite group of American manufacturers, along with the Zippo (lighters) and Harley-Davidson (motorbike) companies, who still market the same product they made nearly fifty years ago. Like the Harley-Davidson "soft tail" motorbike, the famous electric guitars—the Fender Strat, the Gibson Les Paul, the Flying V, and the Rickenbacker 360—have a resonance in popular culture that goes well beyond their identities as manufactured goods. An advertisement for Stickley Furniture features a Flying V propped up against one of the company's products under the title "Uniquely American." The Strat and the

Harley have so much power as images that they stand for a lot more than a means of transportation or a musical instrument—they have become icons for certain masculine depictions of the American way of life: powerful, shiny, complex, rebellious, and eminently desirable.

In an era crammed with every possible technological device and full of manmade noise, the electric guitar has made itself heard. Instantly recognizable and constantly evocative, the output of the electric guitar is recognized as the sound of modern America. Subsequent chapters show how its origins can be found in the multiculturalism of the United States in the twentieth century and how it emerged from a social context of urbanization and the intermingling of races. Its sound reflected the diversity of the United States during a century of great social change, summing up all the conflicting tensions and aspirations of a great nation in one artifact.

Although the focus is on the guitar as a technology, our book is not intended as an encyclopedic overview of all the different American guitars; several other books expertly supply such details. Nor is it intended to be a scholarly monograph that addresses one significant issue in music or society. This purpose too has been served by excellent books based on exacting and detailed research. The aim of this project is to investigate, in an interdisciplinary manner, some of the meanings and functions of this technology. This is not meant to be a picture book that recounts the history of beautiful objects. It is instead a collection of snapshots, each taken from a different aspect, and sometimes overlapping, that collectively give a three-dimensional impression of a potent and influential technological system. It would take several books to do justice to the impact of the electric guitar on our popular culture. Our slim volume is to be taken as a primer that explores the terrain and points the reader in several interesting directions.

Although scholars have already investigated much of the ground scrutinized in this book, plenty of this material has not been covered. Issues of race and gender have been the focus of several scholarly studies, but there are no full-length works about (for example) the rise of Japanese guitar manufacturers, the concept of a guitar hero, or the technology of sound effects. These important topics are addressed in this volume, but the goal has been to connect them together within a meaningful narrative rather than to cover each in an encyclopedic or exhaustive fashion. *The Electric Guitar* is an exercise in putting these pieces together and trying to make sense of the big picture, a picture that draws on innovation, musical styles, electronic technology, and the appeal of rock 'n' roll celebrity. The information presented in this book might not be totally new and original, but its amalgamation of different material and diverse approaches has never before been attempted. It sits at the confluence of technology and culture.

The participants in this project come from backgrounds in history, economics, music, and the social sciences. One of our contributors is a professional musician who uses the electric guitar as a tool of his trade. The approach is interdisciplinary but not theoretical. This is a consequence of the hopeless ignorance of the editor and the need to make this book accessible to the general reader. To this end the use of notes has been kept to a minimum.

Our book is aimed at the sort of person who attended the original Smithsonian symposium—interested in the guitar more as a musical instrument than a scholarly construct, aware of its history and significance but not in a position to put it into a technological or musical context. We hope to gently take the reader into all the territory colonized by the electric guitar and its sound without lingering long enough in one place to spoil the ride.

Sources

Cantwell, Robert. *When We Were Good: The Folk Revival.* Cambridge: Harvard University Press, 1996.

Guitar and *Guitar Player* magazines.

Notes

1. *Atlanta Journal-Constitution,* 15 October 1999, p. R6.

2. Brownie McGhee, interview by Rebecca McSwain, 3 December 1992; Michael Lydon and Ellen Mandel, *Boogie Lightning: How the Music Became Electric* (New York: Da Capo, 1974), p. 26.

3. Daniel Boorstin, "Welcoming Remarks," in *Ethnic Recordings in America: A Neglected Heritage* (Washington, D.C.: Library of Congress, 1982), p. xii.

4. Roddy Doyle, *The Commitments* (New York: Penguin/Viking, 1995), p. 137.

5. Thomas Frank, *The Conquest of Cool: Business Culture, Counterculture, and the Rise of Hip Consumerism* (Chicago: University of Chicago Press, 1997), p. 5.

6. *Los Angeles Times,* 16 August 2000, p. F3.

1 the music

the electric guitar in the american century

Charles McGovern

Beginning with its commercial introduction in the 1930s, the electric guitar developed into perhaps the most important instrument to shape American popular music in the decades to come. It would be impossible to trace American music since 1940 without assigning a central place to the electric guitar. The major forms of American vernacular music that flourished from after World War II through the 1970s—urban blues, country music, rock 'n' roll, and rhythm and blues (R&B)—all depended on the electric guitar for their sound and much of their popularity. Rock 'n' roll, the most widely embraced form of American popular music, could not have existed without the electric, any more than bluegrass music could exist without the five-string banjo, jazz without the saxophone, or European symphonic music without the violin.

The major social changes that remade American life across the twentieth century enabled the rise of the electric guitar and its music. While reshaping American society, migration, industrialization, technologies, markets, and appetites, they also recast American music. The migration of workers from abroad and within the United States, a hunger in these resettled peoples for music that addressed their experiences in their own vernacular, and a cultural climate where self-expression was equated with freedom—all shaped the rise and acceptance of the electric guitar in American life. This chapter explores some connections between American social history and the electric guitar, between guitarists and their music in the years of "the American Century."

multicultural history

The forces that changed American society also nurtured the music and the audience for the electric guitar. The composition of America was altered significantly through successive waves of immigration and urbanization. The great diaspora around 1900 brought to America millions of foreign-born people, chiefly from southern and eastern Europe. This infusion remade urban life and transformed American culture, from language to the popular arts. The newcomers arrived in waves from the late nineteenth century through the mid 1920s, when with the 1924 Johnson-Reed Act Congress all but closed the U.S. borders. During this era more than 10 million new people settled in the United States, and by 1924 more than 13 million people in the country were foreign-born.[1] They became the backbone of the urban industrial working class, settling in as unskilled laborers and factory operatives in the industries that made America the world's strongest economy in the decade after World War I.

They also became the audience for, and often the creators of, the new commercial popular culture that would capture national attention and change the ways people entertained themselves. From the late nineteenth century, immigrants and the city became synonymous with the new forms of popular culture: vaudeville, the nickelodeon and movies, professional sports, beer and dance halls, and amusement parks. These cultural forms took hold throughout America despite an undercurrent of suspicion and xenophobia that frequently erupted into violence aimed at foreigners. Such fears of ethnic, religious, and racial difference did not halt the acceptance of these cultural forms. Ultimately popular culture grew into a billion-dollar business even as it has remained a controversial target of periodic opposition from cultural conservatives, old-line religious leaders, and middle-class guardians of gentility. Nowhere did the migrants have a greater impact than on American popular music. The turn-of-the-century immigrants (in concert with an earlier generation of Germans, Jews, and Irish people) gave American popular music its classic iteration in the Tin Pan Alley song. The New York–based composing and publishing business, known as Tin Pan Alley, popularized "standards," which were generally love songs and novelties mostly in thirty-two-bar form.[2] European migrants and their children made the Tin Pan Alley song central to a host of cultural forms, including the movies, Broadway, radio, the record business, and social dance. Along with the African Americans who created jazz and blues, they made the popular standard into the defining musical form of the first half of the twentieth century.

A second wave of migrations also remade American cities and culture. Millions of rural black and white southerners left the South in the decades after World War I. Their movement north and west decisively reshaped American culture, politics, and mores. The interwar years saw the breakup of the southern labor system—sharecropping and tenant farming—that had kept millions of farmers, black and white, in continual debt. Sharecropping kept southern blacks tied to the land in an oppressive system of economic exploitation and peonage, undergirded by racial segregation and terrorism. During the Great Migration southern blacks left their farms in unprecedented numbers, and often at great risk to their own safety, and headed for the industrial cities in the North. By World War II there were substantial black communities established in virtually every industrial center. The Great Depression, along with New Deal policies that favored agribusiness, accelerated the breakup of sharecropping and set millions more Americans in search of better living farther up the road. During the 1930s white families left the South in massive numbers. The buildup in manufacturing for World War II boosted demand for industrial labor. Even more southerners left their homes for New York, Detroit, Cleveland, St. Louis, Chicago, Memphis, Pittsburgh, Norfolk, Los Angeles, the Bay Area, and Seattle, among others. These migrations continued after the war. By the 1960s more than 10 million southerners had left the region.

This migration had far-reaching consequences. After World War II American culture became increasingly inflected with a southern accent. Music especially took on the voices, stories, concerns, rhythms and rhymes of southern people. Sounds and songs once dismissed as regional and debased now became the bedrock of styles of popular expression, thanks to the intermingling of southerners throughout American cities. Many of rock 'n' roll's most charismatic voices Chuck Berry, Elvis Presley, Fats Domino, Little Richard, Carl Perkins, Jerry Lee Lewis—were unmistakably southern. The fast-talking, rhyming hipster styles of early rock 'n' roll disc jockeys borrowed heavily from postwar southern-born black and white record spinners (whose own styles derived from both preachers and tent show pitchmen) such as Maurice "Hot Rod" Hulburt, Vernon Winslow, Dewey Phillips, and Gene Nobles. The electric guitar's trebly twang and its bent, slurred notes that often hovered between traditional pitches came from vernacular southern traditions of blues and country pickers. As guitar-driven rock 'n' roll, R&B, and even country music occupied the mainstream and became pop music, Americans all over the country embraced southern slang and styles. The European and then southern migrants created and responded to a popular culture that addressed their concerns, featured their heroes, and spoke their language in meaningful ways.

The popular culture industries were the second important element in the electric guitar's rise to prominence. From the turn of the century both large-scale enterprises and small businesses flourished by creating commercial entertainment and amusements. The record business, professional sports teams, amusements and resorts, films, radio, comics, mass-circulation magazines, and television all became large national industries. Americans spent billions of dollars for entertainment and diversion. Workers and their families were integral to these new leisure industries; their passion for diversion became a hallmark of the consumer culture that characterized American life.

Commercial popular culture was more than just a lucrative business. Widespread pursuit of leisure meant that American culture's ethic of self-cultivation and self-expression—the emphasis upon forming and recreating the self through work, self-directed education, and social interaction—increasingly operated not only through work or religion but also by play and diversion. In twentieth-century consumer culture Americans learned how to be Americans increasingly through entertainment. The major culture industries—Hollywood, magazine publishers, the broadcast networks, and the large record companies—most often targeted a broad middling audience with materials that often denied, elided, or diluted the specific experience of class, ethnicity, region, or race. In this way they left the field open for other small businesses who seized an opportunity to serve the underserved markets—those migrants and their communities whose class, color, ethnicity, and accents marked and excluded them from the economic and political mainstream. For many of these workers, popular culture was also a means to claim their cultural citizenship as Americans. Listening to popular music, dancing, following sports teams, attending the movies, and listening to the radio all provided much-needed diversion but also allowed them to become part of, and view themselves as participants in, a specifically American culture. Of course popular culture neither healed nor hid racial and class divisions, but it offered participants of different backgrounds a limited common ground where their passions and talents made them in some measure (and in some circumstances) equals. In that sense commercial pop culture helped working people claim citizenship when commerce, law, or custom otherwise denied them full and equal status. Cultural expression ultimately had political and social consequences, especially for those southern migrants whose music embraced the electric guitar. Through cultural symbols many southerners, especially African

Americans, staked their claims to respect even as American politics resisted their demands for civil rights or economic justice.

The rural migrants brought strong musical traditions to the city in the form of country, gospel, and blues music. In the city, perhaps for the first time, they encountered other forms of music already rooted there—jazz, urban blues ("horn tootin music" in the words of at least one hillbilly band), and the popular fare of Tin Pan Alley, Broadway, and Hollywood.[3] Ensconced in new places, often in unfamiliar living situations with different customs, the new urban settlers mixed their traditions with those of their settled neighbors. The result was a heady new mixture of styles and whole new genres. Out of these migrations and intermingling came such types of music as electric blues, honky-tonk and hillbilly boogie music, R&B, rock 'n' roll, soul music, and much more. Just as the acoustic guitar was a focal point of country and blues music before World War II, in the postwar city the electric laid the foundation for R&B and rock 'n' roll. Country and blues music was also remade by the electric guitar's new sounds. The electric emerged front and center in music geared to listeners and dancers who demanded lyrics that spoke to their experience, rhythms that matched the tempo of their lives, and sounds that resonated with their feelings, reflected their spirits, and stirred their imaginations.

Beyond its roots in working-class and southern lives, the electric guitar's history is explicitly and inescapably multicultural. Different racial, ethnic, and cultural groups have specifically shaped the instrument's history, from its design and production to its repertoire. Taken together, these influences reveal a complex social evolution; the electric's history brings together major strands of American racial and cultural life. The earliest social roots of the electric are Asian-Pacific. In the Hawaiian Islands indigenous peoples adapted the standard guitar from Spanish settlers and used a slide to create the mellifluous and wailing sounds we associate with the steel guitar. On the mainland popular culture reflected a fascination with Hawaii after the turn of the century.[4] The instrumental textures and tones of Hawaiian music, particularly the ukulele and the slide guitar, made their way into the popular marketplace in the early decades of the century, and Hawaiian sounds fascinated the early generations of steel guitar players in country music.

The technique of using a slide on the strings had a tremendous influence upon black and white musicians in rural areas, where it became a bedrock of rural sounds. African American musicians had an indelible impact on the electric's evolution through their widespread adoption of both slide and resonator guitars (which were also identified with Hawaiian-style music) from their introduction in 1925. Black guitarists brought the slide to prominence in blues music in the 1920s. Such stylists as Sylvester Weaver and Blind Willie Johnson contributed to the blues emphasis on bent and frayed notes, microtones, and vibrato, played with a slide and by bending and stretching fretted notes. Musicologists and others have long observed that such sounds have both antecedents and contemporary analogs in African music, especially in the western part of the continent, where many of the slaves came from. The African American prominence in developing the vocabulary of guitar expressions follows from these shared traditions. Sylvester Weaver's "Guitar Blues" gained fame as *the* single piece for budding steel guitarists to learn when it was popularized as "Steel Guitar Rag" by Leon McAuliffe with Bob Wills and the Texas Playboys. After World War II blues players led by Muddy Waters and Elmore James, and later rock guitarists, adopted the slide's arsenal of wailing, crying, and screaming sounds.

Both the fluid sound of the bottleneck slide and the bending of notes between standard pitches are basic to blues guitar. When guitarists began adopting electricity in the 1930s, these two techniques found increasing use as African American guitarists in particular experimented with the sustain, distortion, and vocalisms that amplification offered. Aaron "T-Bone" Walker and Charlie Christian, the acknowledged founders of electric blues and jazz guitar, drew freely on these vocabularies, even if others who followed, such as Albert Collins, Pete Lewis, John Lee Hooker, Pat Hare, and many others, explored the limits of bending, microtones, and distortion much more fully. From the outset these techniques were at the heart of African American electric styles. Walker bent, pushed, and pulled notes continually in his playing. On his signature piece "T-Bone Boogie" (1945) he coaxes subtly shaded quarter- and halftones, bending just under the pitch and constructing lines that slash through rhythms and chord changes. Teddy Bunn, an undervalued guitarist who straddled the commercial lines between blues and jazz throughout his career, turned in a virtuoso performance on Edgar Hayes's "Fat Meat and Greens" (1949). His solo was an almost continuous series of bent, slurred, elided, and staccato microtones, laced with distortion. Bunn offered a primer of blues and rock 'n' roll licks on record long before the full emergence of Chicago electric blues guitar or the better-known antecedents of rock 'n' roll, such as Jackie Brenston's "Rocket 88" or Big Mama Thornton's original "Hound Dog." Christian and Walker both used distortion sparingly, as just another element in their vocabulary of riffs and approaches, but they both established that a distorted sound, heavy with overtones and static, could be a viable part of the electric's overall aesthetic.

Latino peoples in North America influenced the electric's evolution as well. Latinos in the Southwest shaped the acoustic guitar's basic sounds, adopting strums and tunings handed down by descendants of Spanish settlers. They influenced both R&B and rock 'n' roll music almost from the first. Americanized Latin rhythms such as the mambo and the cha-cha became staples of R&B tunes, while more importantly, Latin tunes and techniques became basic fare for aspiring electric players. Such tunes as Ernesto Lecuona's "Malaguena" and Alberto Dominguez's "Perfidia" became standards for aspiring electric guitarists, staples in instrumental bands playing what we now call surf music. Latino musicians were active in rock 'n' roll, country, and western swing bands especially in the western, southwestern, and northeastern United States.[5] Long before Carlos Santana emerged as one of the most important and visionary guitarists working in popular music, Latinos had shaped the history of the electric. They were also important in manufacturing electric guitars, especially at Fender, where they were involved in the critical operations of pickup winding and wiring, along with body finishing. In other words, while the designs were the product of Leo Fender with his key team of Freddie Tavares (himself Hawaiian), George Fullerton, and Forrest White, the execution of the *look and sound* of Fender guitars owed a great deal to Latinos; the classic Fender appearance and tone were products of Latino hands.[6]

European Americans were predominant in the electric guitar's history as designers, businessmen, and consumers. Yet even here specific ethnic traditions helped shape the electric guitar's evolution. The preeminent luthier John D'Angelico drew upon Italian traditions of violin making, while Elmer Stromberg brought Swedish cabinetry techniques to his guitars. John Dopyera, inventor of the resonator guitar, was an immigrant Czech violin maker. Working with brothers Rudolph and Emil, he wedded his instrument-making techniques with modern machine tooling to design and make the National and Dobro guitars, the loudest acoustic guitars on the market before the perfection of electric pickups. Ethnic traditions also permeated the styles of influential guitarists. Eddie Lang (born Salvatore Massaro) died before adopting the electric guitar, yet his fiery playing and precise articulation—recalling the trills and plectrum style of Italian mandolinists—influenced such important early jazz guitarists as Django Reinhardt, Charlie Christian, and George Barnes. Reinhardt, the most influential guitarist in jazz before Christian, built his style upon the melodies and dances of his Gypsy people, the Manouche. Such solos as "Finesse" and "Improvisation" show his Gypsy roots as well as his jazz conception. Southwestern white electric guitarists such as Eldon Shamblin, Porky Freeman, Herb Ellis, and Barney Kessel incorporated the German and Czech music that were staples of Texas and Oklahoma dances, as well as jazz and swing sounds, into their styles. For example, Freeman's solo on Jack Guthrie's original version of

"Oklahoma Hills" features trills, glides, and rapid-fire descending notes that recall polkas as well as jazz.

Women have been involved in the electric's history from the first, although their roles have been overlooked. Certainly the instrument has been overwhelmingly masculine in its imagery as well as in terms of its practitioners. Few women came to any public prominence playing electrics before the 1970s, yet women were involved with every phase of the instrument's history. Women had always been identified with the guitar (as well as the banjo and piano) as exemplars of the genteel and civilizing aspect of music for the middle class. Images of women with guitars commonly adorned advertisements, associating the guitar with refinement. With the rise of the electric, women were often pictured playing lap steel guitars in advertisements and on radio. Author and guitar expert Tom Wheeler dates the earliest public image of an electric solidbody Spanish guitar to a 1935 publicity photograph of the Sweethearts of the Air.[7] Women were also involved in production and manufacture, often working in guitar factories doing wiring and detail work.

As musicians, women were innovators, even if they were few in number. Most prominently Mother Maybelle Carter practically invented rhythm guitar as we know it in country and later rock 'n' roll music. Her trademark brush strokes and rhythmic patterns (heard on signature tunes such as "Wildwood Flower" and "Keep on the Sunny Side") were adopted by millions of musicians. Gospel singer Sister Rosetta Tharpe was an influential figure in midcentury music, fusing sacred and secular traditions in her blues and jazz-tinged songs and playing. She featured her own guitar, first a resophonic National and by the 1940s an electric, on virtually all her recordings. Her explosive picking and propulsive strumming, heard to great effect on the 1946 "Strange Things Happening Every Day," set her records apart. Tharpe's singing and picking influenced a whole generation of founding rock 'n' roll musicians, including Carl Perkins, Sleepy LaBeef, Jerry Lee Lewis, and Johnny Cash. Other guitarists, such as Memphis Minnie, Mary Deloatch, Mary Osborne, Martha Carson, and Mary Ford, recorded and performed with electrics in a variety of styles in the years before rock 'n' roll emerged.

Yet by and large women were less visible as instrumentalists, particularly on electric Spanish guitars, which were associated with masculinity especially after the emergence of rock 'n' roll. While it was acceptable for women to be seen playing soft music with domestic themes on comparatively genteel acoustics or Hawaiian lap steels, there were few women like Tharpe or Mary Osborne soloing on electric or, like Mary Ford, performing to packed houses and on television. Cordell Jackson, an early exponent of the electric, recalled that when she appeared on local radio in Mississippi and Memphis, male performers were

Cordell Jackson was an early exponent of the electric guitar. She faced opposition from men because of her unrestrained playing. Courtesy of Cordell Jackson.

irritated and threatened by her fast strumming and picking. "No man could touch what I was doing," she recalled with no little pride. They repeatedly discouraged her or tried to take the guitar away from her.[8] Often male musicians would accept female band-mates in the role of singer, rhythm guitarist, or bassist (instrumental roles that lent fundamental support but did not take the spotlight), but having a woman claiming a place for herself in the ensemble with an electric challenged traditional gender roles too forcefully. Even rockabilly singer Wanda Jackson played acoustic rhythm guitar when she emerged in the 1950s, although within a decade she was performing on an electric.

electrifying the sound

Cordell Jackson's experience gets to the heart of the electric's appeal. It gave its users not only more sonic and tonal options, but a first-time ability to be heard. That new autonomy and freedom transferred from instrument to owner; the amplified guitar conferred autonomy, confidence, and power. The multicultural history of the electric shows us that power—volume, presence, the ability to be heard relative to other instruments—assumed diverse forms for musicians and listeners. Even as most guitarists sought a better, louder, and often "purer" tone, free from overtones or distortion, their efforts also fostered new possibilities glimpsed by others who followed or listened to them. In other words, the electric became a symbol as well as a working instrument, a visual and aural force in American culture. As a cultural symbol the sound and style of electrics allowed musicians and listeners alike to imagine new possibilities and make claims for themselves that resounded far beyond the dance hall, beer joint, or school gym. Through its multicultural appeal and its centrality to working people's music, the electric guitar produced sounds and gave voice to impulses that would transform American life.

The symbolic and subjective aspects of the electric were rooted in the same social transformation that remade America—the migration to the city and the experience of new technologies of mass communication and transportation. When the country got to the city, it plugged in. The relatively louder volume required in urban and more populous nightspots created the functional need for electric guitars, the ability to be heard and to put forth a volume sufficient to entertain or propel large crowds of dancers across the floor. The first generation of players and listeners who encountered the instrument in the late 1930s through the 1950s did not necessarily take electricity for granted. Many, such as Riley B. B. King or Carl Perkins, grew up on farms with no electricity. Yet part of the fascination with electrics was also the mystique of electricity itself, which had long held profound meanings for rural people, even before they moved from country to city. Genuine popular wonder and curiosity, along with energetic public relations efforts of Thomas Edison and others, had conferred upon electricity not only a benediction of progress or convenience but the aura of the otherworldly. Electrical power meant transformation, a connection to the cosmos, a vital tie to the future, and the attainment of strength and power, most often the power of God. Whether in communications (telephone and radio), or in the transformation of the environment (electric light's manifestation of artificial daylight), or in the tremendous extension of human capabilities (through motors, generators, and other machinery), electricity assumed a power of almost mythic proportions for many Americans. Electricity itself meant a momentous change: in farm labor, in daily life, in the ways people worked, communicated, and thought. It offered not only the symbolism of heavenly power but transcendence of human capacity beyond the imagination. The lightning bolts that adorned the Epiphone Electar and K&L amplifiers and Rickenbacher and Supro guitar nameplates in the 1930s immediately symbolized this power, implying that the guitarist had in hand something beyond a mere musical instrument. When a 1936 Rickenbacher brochure enthused, "Brother musician, listen to a miracle!" it made clear that the electric guitar would confound the expectations and transform the ways people played.[9] For the musicians who embraced the early electric, that miracle meant being heard and expressing themselves beyond their wildest imaginations.

Electrics reoriented the business of booking live music and altered the composition of bands. An electrified ensemble could replace a bigger band, a fact that club owners (always looking to get more musical bang for less buck) quickly turned to their advantage. It is no accident that the electric's rise in blues and jump music coincided with the decline of the big band or that so many of these influential small bands, such as Louis Jordan's Tympany Five, Roy Milton's Solid Senders, T-Bone Walker's group, and the King Cole Trio, all prominently featured the new guitars. The postwar downsizing in many western swing bands reflected this new reality; the horn players departed, but the electrics, steels, and amplified mandolins grew more prominent in the postwar ensembles of Bob Wills, Hank Penny, Hank Thompson, and others.

The electric guitar changed perceptions and challenged expectations for both musicians and listeners. It created fissure lines between musicians and conflict in the audience. The successful adoption of pickups in National and Rickenbacher guitars, along with growing sales of add-on pickups made by DeArmond and other manufacturers, had established the electric guitar as more than a novelty by the late 1930s. Yet playing electric hardly guaranteed aesthetic or commercial success for aspiring musicians. The electric in fact provoked audiences, critics, and musicians, many of whom reacted harshly to the new sound. Frank Victor, an early electric advocate and a *Metronome* columnist, told guitarists in 1936 that the all-powerful Local 802, the New York City chapter of the American Federation of Musicians, disapproved of electrics. This distrust made it difficult for guitarists to get work playing the electric, even as some of the first jazz and pop artists were beginning to experiment with it on record. At that time Les Paul came to New York to work with vocalist Jim Atkins and bassist Ernie Newton in Fred Waring's big band. Years later Paul remembered that his electric prompted scores of letters from Waring's audience protesting the "way-out" sound. Many demanded Paul's dismissal from the band, or at the very least silencing the electric. The electric also prompted scorn from many of Paul's fellow musicians in the band. But Paul also remembered an outpouring of support from other listeners and musicians.[10] Even so, the controversy indicated how troubling the electric was for so many listeners who deemed the new guitar's sound otherworldly, strange, or abrasive. George D. Hay, the "Solemn Old Judge" who ran the *Grand Ole Opry* program from Nashville's WSM radio station, detested electrics and generally forbade them (along with drums) in the country music mecca until after World War II, when they made inroads into that traditionalist stronghold.

Blues guitarists encountered similar opposition, and it took years for the electric to gain full acceptance in that genre. Even before T-Bone Walker's stunning national debut on records in the early 1940s, the famous blues pickers Big Bill Broonzy, Tampa Red, and Lonnie Johnson all recorded or played with electrics. Yet only in the late 1940s would the electric gain sufficient acceptance to become standard. When label owner Leonard Chess allowed Muddy Waters to update his traditional (and lucrative) acoustic sound on record with an electric guitar in 1950, the popular response was immediately enthusiastic. But Chess had fought Muddy's innovations. He disliked the electric's sound and feared that it would alienate Waters' loyal listeners who preferred the down-home sound—a lone guitar with a slide, itself a holdover from the Delta—still popular among Chicago immigrants. Chess did not allow him to record with a full electric band until long after he had established it as his sound in Chicago clubs.[11] The electric challenged preconceptions not only about the guitar but also about music in general, even as musicians such as Paul and Waters were engaged in practices that would radically alter ideas of what was possible for guitars.

Most of all, electrifying the guitar allowed musicians a greater range of choices in developing their own individual sounds or voices. The timbre and tonality of the electric intrigued and challenged guitarists, and with its rise an entire subculture of musicians and businessmen created products to make new sounds or modify existing ones. From early add-on pickups sold by Volu-Tone and DeArmond, to breakthrough postwar technologies—humbucking pickups, reverb, and tremolo—to the latest effects and digital sampling boxes today, electrifying the guitar took musicians into new terrains of technology. Guitarists came to view the creation of musical sounds as not only a collaborative process between musician and instrument, but one in which external equipment took on an increasingly significant role. Yet at any level of aspiration or accomplishment, playing with power meant empowerment. The electric guitar allowed you to be heard by all, and you could rely on an increasingly vast array of tools to help shape your voice.

The pickup not only enlarged the guitarist's sound and range; it also inevitably changed the ways in which players approached the instrument, and indeed it changed the fundamental textures and sounds of guitar music itself. Jazz and pop, which had long been dominated by horns and the piano, saw little change, but where the guitar was already central—blues and country—the electric transformed the music. In any setting an electric guitar could become a solo instrument. Once limited to the rhythmic underpinning, the electric guitar now stepped forward as an equal among horns, fiddles, pianos, and voices. Electrics worked easily in dialogue with other sounds and at times dominated the music entirely. The bending of notes and microtones, already at the heart of both blues and country music, assumed a greater importance simply through the greater volume. The electric could sustain as beautifully as a fiddle, or moan along with the mournful voice of the blues, or punctuate the vocals with the same assertiveness as a horn. These qualities encouraged guitarists to explore ideas and phrasings associated with different instruments—saxophones, trombones and trumpets, pianos, and drums.

In the first years of electrification, guitarists took these possibilities in different directions. Jazz guitarists generally downplayed the sonic possibilities of overtones, distortion, and feedback. These sounds constantly cropped up in the daily lives of electric guitarists as many pickups and amps performed inconsistently, often delivering feedback and static or cutting out at unpredictable moments. Instead jazz players emphasized hornlike lines that were fluid and smooth, that allowed them to emulate the legato phrasing of a soloing tenor saxophone. Charlie Christian became known for his articulate driving and inventive lines and riffs. He relied on the sustain of his Gibson ES-150 to construct long flowing phrases and to serve as another "horn" with the clarinet, tenor sax, and trumpet in the Benny Goodman small groups. Their complex arrangements utilized Christian as a voice in the harmonized ensemble passages, and only when he soloed would the listener realize he was a guitarist. He downplayed the use of vibrato or distortion, but his lines were steeped in blues inflections, as in such tunes as "Breakfast Feud" and "Shivers." His solos on the several extant takes of "Breakfast Feud" show Christian's many techniques. He built long hornlike lines that were unprecedented. He bent, slurred, and slid into notes throughout these lines, alternating them with complex, precision picking. While he did not probe the range of sonic possibilities, at the time of his death in 1942 (at age twenty-five) he had created a palette of tones and techniques that set the stage for jazz guitar for the next generation.

The cohort of jazz players that emerged after Christian in the 1940s and 1950s explored the guitar's connections to horns, piano, and rhythm. They created distinctive sounds within a rather narrow range of technical options and standardized equipment. Jimmy Raney, John Collins, Tal Farlow, Barney Kessel, Kenny Burrell, Jim Hall, Johnny Smith, Herb Ellis, and Wes Montgomery all worked with essentially the same resources—electrified hollowbody guitars and tube amps—yet they all found distinctive voices as well as styles. Each explored long horn-inspired melodies while producing sounds that eschewed distortion and overtones. Wes Montgomery's trademark warm sound, which came to prominence in the late 1950s, resulted in part from using his thumb instead of a pick. His stunning voicing of octaves and chord melodies on such tunes as "Four on Six" and "Too Late Now" recalled pianists such as Errol Garner and Milt Buckner, while he often turned the guitar into a percussion instrument, as in his 1963 recording of John Coltrane's "Impressions." Johnny Smith, influenced by years as a trumpeter and as a sight-reading musician in the NBC studio orchestra, voiced his chord melody style very much in the manner of a pianist, as heard in his classic 1952 recordings of "Moonlight in Vermont" and "Tenderly." His quintet arrangements stressed the hornlike midrange of the guitar, as in such recordings as "Taboo" and "Sometimes I'm Happy" from the mid 1950s. Smith, like Christian before him, created an ensemble sound that made the electric seem like a saxophone.

From the mid 1930s through the mid 1940s, blues and country guitarists began exploring the guitar's sonic possibilities. Country guitarists often crossed genres and confounded audience expectations by mixing sophisticated jazz harmonies with a whining treble sound that ripped through their ensembles. They often played off the tensions between simple melodies and complex approaches to harmony, using both dissonance and slurred notes. Electric steel guitar pioneer Bob Dunn's jazz-inflected playing with Milton Brown offered great swoops and glides, dissonances and vocalisms that took the steel far from the sweet Hawaiian traditions where it was rooted. As early as 1935 Dunn explored dirty tonality and distortion to an extent that few would approach till over a decade later.[12] His work on "You're Bound to Look like a Monkey" and "Taking Off" especially showcases this approach. Lester "Junior" Barnard of the Bob Wills band became famous in the early 1940s for his bluesy and distorted sound, full of bent notes and chords that were fluid and funky. Like Dunn, Barnard united jazz improvisation with an approach to guitar that produced distorted, crude, and primitive sounds. His style contrasted with the nimble acrobatics of such contemporaries (and veterans of Bob Wills's band) as Eldon Shamblin, Jimmy Wyble, and Cameron Hill. Numerous country guitarists played off the tension between biting trebles and sonorous sweet sustain. Faye "Smitty" Smith's solos in the early 1940s with Ernest Tubb's band were rudimentary but influential. Guitarists all over the country were soon imitating his piercing treble, and Tubb's later guitarists Billy Byrd and Leon Rhodes found themselves playing with a similar tonality to emulate the trademark Tubb whine. Such early steel guitarists in western swing as Noel Boggs, Joaquin Murphy, and Leon McAuliffe all aimed for sounds that were both piercing and mellow. The steel, like the electric Spanish guitar, cut through the sound with a mixture of biting treble and sustain that could be alternately soothing, haunting, and rousing. In the mid 1940s, country guitarists such as Porky Freeman and Merle Travis produced sounds that combined the intricacies of jazz horn with the slightly muffled "chicken pickin'" style of rural guitarists both black and white. When steel guitarist Speedy West and his partner Jimmy Bryant (on electric Spanish guitar) gained public prominence in the early 1950s, they played astonishingly intricate ensemble passages, such as in "Stratosphere Boogie," that seemed to bridge the gaps between the farm, the nightclub, and outer space. Bryant and West perhaps epitomized the contrasts of tone, texture, and harmony in country guitar. Sharp-edged piercing sounds and precision statements of complex harmonies marked their work. They incorporated blues, jazz, and country in their playing, and while it was seen as country, it pointed the way toward the recombinant music we know as rock 'n' roll.

Blues guitarists also experimented with tonality, distortion, and overtones—less blending in with other instruments than contrasting with them. T-Bone Walker's phrasing, built on his own experiences as a dancer, became influential as much for its forceful rhythmic cadences as for the way he seemed literally to push the notes right through his speaker. In "T-Bone Jumps Again" (1947), for example, his solo evokes a tap routine in its rhythms and accents, even as he bends and slides notes, teasing the listener with an almost infinite arsenal of colorations of the same note. In "Call It Stormy Monday" Walker alternated jazzy chords with biting bent-string lines, but whereas Charlie Christian's guitar blended into the instrumentation of Goodman's groups, Walker's guitar stood out in sharp relief against his group sound. His piercing attack and tone worked against the instrumental grain, calling more attention to the wide array of slurs, bends, and colorations in his playing.

Guitarists who followed Walker soon began building upon his innovations while borrowing his phrases. Appropriation of T-Bone's licks continued well into the 1960s, from postwar bluesmen such as Pee Wee Crayton and Lowell Fulson to founding rock 'n' roller Chuck Berry and, through him, aspiring guitarists around the globe. When he emerged on record in the early 1950s, B. B. King adopted his string-bending vibrato from Walker. King imitated the bottleneck slides of country blues, along with the gently wavering slurs of the steel guitar, to create his own style. King's approach emphasized almost continuous vibrato, along with a call-and-response interplay with his own vocals. These two hallmarks of his playing influenced countless younger guitarists for many decades. His breakthrough hit "3 o'Clock Blues" (1953) featured his trademark sound along with the tightly squeezed bends and slurs and the short choppy phrases that characterized his work.

While Walker seldom used distortion as a trademark feature of his style, it cropped up as a form of accent, another coloration of notes and tones to convey his stories. By the early 1950s other guitarists, many of them from the Mississippi Delta region, began a serious exploration of sustain and distortion and built it into their sound. Pat Hare, who played with a number of West Memphis bluesmen such as Junior Parker and James Cotton, exhibited a tone so distorted and tortured that it sounded like heavy machinery, menacing and compelling. Heard most memorably on James Cotton's 1954 "Cotton Crop Blues," Hare created a treble-heavy distortion that nearly overwhelmed his tonality, the notes scarcely audible through the static, scratching, and overtones. Clarksdale, Mississippi, native Izear "Ike" Turner was one of the most accomplished and influential of the electric Delta guitarists. After his early breakthrough on record with "Rocket 88" in 1951 (a driving boogie featuring his own feisty piano), Turner concentrated on the guitar. In the mid 1950s he explored his Fender Stratocaster's capacities for sustain, reverb, and tremolo to the utmost. On those recordings Turner produced slashing and soaring solos that even fifty years later seem otherworldly. On the 1957 recording "Much Later" Turner's guitar gouges the melody and cuts through his group's rhythmic background, tearing through the insistent horn riffing in the song.

Perhaps most intriguingly, Bo Diddley, born Otha Ellas Bates McDaniels in McComb, Mississippi, brought a percussive approach to his guitar that subordinated all aspects of his playing to the rhythm. Moreover, he consistently manufactured new sounds for himself, constructing his own guitars and sound equipment including amplifiers and sound effects. His playing combined a characteristic use of heavy reverb and distortion along with a restless sonic appetite. Sometimes his playing emulated the roars of the jet engines whose images adorned his instruments; at other times he created an atmospheric sound featuring layers of reverb and echo. On his tour de force "Mumblin' Guitar" Diddley took his guitar through its paces, generating sounds reminiscent of a turbine, bongos, and an electric percolator. If the jazz guitarists showed the possibilities of playing like horns, the electric bluesmen used the guitar to imitate elements of the natural and the industrial world. Their music echoed the journey that so many took from country to city.

recombinant music

The music we know as rock 'n' roll featured electric guitars prominently. Built upon all these styles—blues, country, R&B, gospel, pop, and jazz—rock 'n' roll was essentially recombinant music. Its strength and appeal resulted from the breadth of its roots and its fresh reassembly of those elements. Rock 'n' roll music saw guitarists expanding the sonic palette of electrics, even as they explored the instrument's capacities for both rhythm and solo playing. Chuck Berry's style, rooted in T-Bone Walker and country music, propelled his songs forward rhythmically, first and foremost. His chord techniques adapted the boogie-woogie piano figure to the guitar, and his solos always carried the beat, even more than the melody. In "Too Much Monkey Business," "Maybellene," and "Johnny B. Goode," for example, Berry stretches the contours of chords and melodies, bending his chords and hammering them in rhythmic figures. In these records and others he laid the groundwork for most of rock 'n' roll rhythm guitar as well as lead!

Memphis legend Steve Cropper, Louisiana native James Burton, and Elvis Presley's great accompanist Scotty Moore all developed influential styles that relied heavily on the rhythmic aspects of the electric. But instead of the four-to-a-bar accents of great acoustic rhythm players such as Freddie Green or Maybelle Carter, the electric guitar's role was to supply heavily accented figures that complemented the four-beat rhythms carried by drums and bass. The great rock 'n' roll guitarists of the 1950s heavily accented the rhythm. Danny Cedrone's solo on

Mickey and Sylvia. Sylvia Robinson might have been playing second fiddle to the highly regarded guitarist Mickey Baker in this popular R&B duo, but she went on to be the brains behind some of the pioneer rap recordings. Courtesy of Jonas Bernholm Collection, Division of Cultural History, National Museum of American History, Behring Center, Smithsonian Institution.

Bill Haley's "Rock around the Clock," Buddy Holly's solo on "Peggy Sue," and Mickey Baker's stinging fills and lead on his "Love Is Strange" (with Sylvia Robinson) all exemplify the ways in which rock 'n' roll guitarists forged together rhythm and lead lines. Not until the end of the decade did a new generation of guitarists make clearer the distinctions between lead and rhythm. Instrumental favorites such as Duane Eddy, Dick Dale, and Frank Virtuoso, along with guitar groups such as the Ventures, the Chantays, the Trashmen, and the Fendermen, brought electric guitar front and center not only within vocal combos but on its own as a featured instrument.

Electric guitar music largely grew by altering and adding together many different styles, techniques, and sounds. As we have seen, the electric became important in genres that are easily categorized, yet the music's history and the musicians' biographies confound our conventional expectations. Electric guitarists often had multiple musical influences that crossed social boundaries. In other words, stylistic designations often obscure the music's social history as lived by the musicians and their communities. Because the guitar had a prominent place in different cultures and communities, its music crossed the social and commercial borders of race, class, ethnicity, and region that were used to map different genres. The merchants who marketed the music—recording companies, publishing houses, promoters, radio networks and stations—assumed these same social barriers in constructing their markets, and to a certain extent these categories still have power today. Ideas of what constituted jazz, blues, and country were in good measure the creations of businesspeople, crafted from existing perceptions of social, cultural, and geographic difference. They existed principally to sell recordings and sheet music, not to designate aesthetic distinctions or social history. While we inherit those categories, we are not bound by them; more importantly, neither were many of the guitarists discussed here.

For electric guitarists, geography, history, and social context as much as talent and aesthetics shaped musical expression and experience. We see commonalities in the music of diverse artists who never occupied the same stage, touring company, or marketing category. Charlie Christian gained fame as a pioneer of electric jazz guitar, yet he shared much with other southwestern guitarists who, strictly speaking, did not play jazz, from his old friend T-Bone Walker to such country guitarists as Eldon Shamblin or Porky Freeman. Guitarists learned the instrument from eclectic sources—the acoustic guitar's multicultural roots and the recordings and radio whose history coincides with the electric's. This is surely not a novel insight, but if we assume this multiplicity instead of continually being surprised by it, the grounds for inquiry into the instrument's history shift. Instead of constantly trying to refine and define the boundaries and traits of a given style, we might ask specifically how players came to their knowledge and what effect their music had, not only upon aesthetics but upon their view of the world.

Some cases illustrate the multiple roots of electric guitar. Les Paul, known as Rhubarb Red in his early career in Wisconsin and Chicago, played a style of country music that was common in the Midwest, yet he longed for and ultimately made a transition to playing jazz.[13] His influences included Django Reinhardt, and like his idol he became known for his rapid runs and clear, trebly sound. Yet his duets with Mary Ford reveal that he remained in many senses a country guitarist; he often laced his music with a good chunk of fiddle tunes and eastern European polka runs thrown in for good measure. B. B. King has always acknowledged his many influences: Bukka White, Lonnie Johnson, Django Reinhardt, and the Hawaiian steel guitar. White and Lonnie Johnson are most easily heard in B.B.'s style, but the presence of Django and steel guitar initially confounds the listener. Trying to imitate the steel guitar's fluid slide, King formulated his arm-bending vibrato, which in turn has been one of the great influences upon the development of modern rock and blues guitar. I suspect that Reinhardt's tone and attack made an imprint on King's style, particularly his ability to control his guitar's volume with his hands and not his amplifier. Godfather of rock 'n' roll guitar Scotty Moore has often told stories of his admiration for Chet Atkins, and during the early years with Elvis he went out of his way to create a sound very much reminiscent of his hero, down to securing a custom-made Ray Butts EchoSonic amp just like Chet's.[14] Commentators have noted the way Scotty freely adapted from the blues records that Sun Records boss Sam Phillips played him—the classic Sun blues sides featuring Pat Hare, Floyd Murphy, and Ike Turner, among others. Yet Scotty stated that Tal Farlow and Django were among his principal influences, along with Josh White. These were the guitarists he emulated, even if he ultimately developed a very different style in which their imprint was less apparent.

On one level this makes plenty of sense. Musicians have ears, and their tastes are as diverse as their fingerprints. From the 1930s onward electric guitarists absorbed an astounding number of different musical styles, distilling them into marvelous rich forms. Yet given the isolation that many of them grew up in, and the many ways American culture was both subtly and blatantly segregated, the wonder is that they did experience a wide variety of musical styles and talents. From the dawn of commercial recording, musical genres presupposed a social base. Country music was largely for and by rural folk (and their cousins recently off to the city), and blues and R&B were largely for African Americans in the country or the city. When rock 'n' roll became popular, its youthful appeal as well as its stylistic idiosyncrasies made it the province of young people. But people were listening across and against the grain of their supposed tastes and social stations at this time. The audience of teens and young adults that followed pioneering R&B disc jockeys such as Daddy-O Daylie in New Orleans, Zenas Sears in Atlanta, Alan Freed in Cleveland, Dewey Phillips in Memphis, Hound Dog Lorenz in Buffalo, Hunter Hancock in Los Angeles, and Gene Nobles and John Richbourg in Nashville were initially African American. This audience grew more integrated by the mid 1950s. So when Elvis Presley, Fats Domino, Chuck

Berry, Little Richard, and others became national celebrities, their audiences were mixed. That mixing threatened the established social order in a largely segregated country, but it represented a reality that had long existed in musical experience. Musicians themselves were inexorably a part of this process. B. B. King and many of his generation had grown up as much on the *Grand Ole Opry* as on the blues. Calvin Newborn, a guitar legend in Memphis and brother of the great pianist Phineas junior, recalled that his first guitar tune was the Bob Wills version of "Steel Guitar Rag."[15] Ike Turner cut that tune as well, while Freddie King paid tribute to western swing with "Remington Ride."

Likewise, many hillbilly guitarists had wide-ranging influences. Porky Freeman, Jimmy Bryant, Grady Martin, Roy Lanham, Jimmy Wyble, and Billy Byrd, along with steel players such as Noel Boggs, Buddy Emmons, and Speedy West—all of them legends in Nashville and West Coast studios—were as much at home with jazz, swing, and even bebop tunes as they were with fiddle tunes like "Arkansas Traveler." Perhaps the most famous exemplar of this crossover from country to jazz is Hank Garland, who made his living in the Nashville studios during the 1950s but who also cultivated a formidable jazz sensibility, which was just blooming when his career was curtailed. Country guitarists who came to prominence in the 1960s like Jerry Reed, Roy Clark, and Buck Owens were rooted as much in R&B as in country. Owens in particular made a number of format-defying recordings that must have driven the salesmen at Capitol Records crazy. His recordings of such rock 'n' roll tunes as "Charlie Brown," "Memphis," and "Twist and Shout" (performed live wearing a Beatles wig) reveal his openness to different styles of music.

Rock 'n' roll's popularity in the late 1950s inspired thousands of young people to play electric guitars. The first generation of rock 'n' roll guitarists drew on sources from every conceivable genre. These musicians and their generation absorbed the wildly distorted sounds of the raw electric blues, along with the snaky twangs of country chicken pickin' and the warm hollowbody sound of jazz players. Ultimately they forged their own sound, but they took their influences from wherever they could find them. Belgian guitarist Toots Thielmans had an impact upon John Lennon, while jazzman Howard Roberts influenced Chicago guitarist Terry Kath. Pops Staples had an impact upon Robbie Robertson (of the Band), and Hungarian-born jazzman Gabor Szabo changed Carlos Santana's playing. The list is endless.

By the rock 'n' roll era, exposure to different forms of music had an indelible impact upon the ways people learned to play and listened to the electric guitar. Radio and records offered an endless array of sounds and styles. In many areas the radio provided an energetic listener with a musical education to last a lifetime. Recordings had the same broadening impact. By the 1960s long-playing albums had entered the youth market, making it even easier for young musicians to hear virtually every style. That eclecticism became even more pronounced later in the decade, when rock guitarists and their peers actively began investigating the roots of their favorite music. We have long known of the British Invasion groups' reverence for Chicago blues; the Rolling Stones, Fleetwood Mac, and John Mayall's Bluesbreakers spearheaded an appreciation of electric blues that revived a number of African American musicians' careers and set the electric blues on a path to worldwide popularity. Such early 1970s groups as Roomful of Blues and Asleep at the Wheel, now institutions in their own right, became living workshops in the history of R&B, jump, and country music. Other groups, whether the Stones or the Ramones, still remained true to their own bedrock influences. Each new style of rock 'n' roll reveals still more complex roots and sources, and the success of compact discs has made it feasible for more musicians today to hear a greater variety of older music than in previous years. Rock 'n' roll grows more recombinant, more diverse, over time.

Yet even as the guitar evolved along diverse social paths, in numerous styles, and from myriad influences, a common thread of experience united most electric guitarists. For all its eclecticism, the electric offered assimilation. As country, blues, jazz, and rock 'n' roll each became synonymous with American culture, the electric guitar symbolized both the individualism and the freedom associated with the United States. The instrument enabled young musicians to extend their capabilities and to be heard. The electric's voice was new as well as powerful. It took relatively little effort—six hours to six weeks—to learn the rudiments of blues, country, or rock guitar. In the postwar years, burgeoning markets and numerous styles meant that there were few rules to absorb and still fewer that were inviolate. For many electric guitarists, that lack of tradition meant freedom to invent themselves. It gave musicians of all backgrounds, capacities, and aspirations a means for singing out loud and strong. In other words, it was cool.

Sources

Butsch, Richard. *The Making of American Audiences: From Stage to Televison, 1770–1990.* New York: Cambridge University Press, 2000.

Nasaw, David. *Going Out: The Rise and Fall of Public Amusements.* New York: Basic, 1993.

Palmer, Robert. *Rock & Roll: An Unruly History.* New York: Harmony Books, 1995.

Peiss, Kathy. *Cheap Amusements: Working Women and Leisure in Turn of the Century New York*. Philadelphia: Temple University Press, 1986.
Rosenzweig, Roy. *Eight Hours for What We Will: Workers and Leisure in an Industrial City*. New York: Cambridge University Press, 1983.
Snyder, Robert. *Voice of the City: Vaudeville and Popular Culture in New York*. New York: Oxford University Press, 1989.

Notes

1. Matthew Frye Jacobson, *Whiteness of a Different Color: European Immigrants and the Alchemy of Race* (Cambridge: Harvard University Press, 1998), p. 43.

2. Russell Sanjek, *American Popular Music and Its Business: The First Four Hundred Years,* vol. 3, *From 1900 to 1984* (New York: Oxford University Press, 1984), pp. 32–44.

3. See, for example, *The Billboard 1944 Music Yearbook* (Cincinnati: Billboard Publishing Co., 1944).

4. On the Hawaiian connections to steel and resonator guitars, see Bob Brozman, *The History and Artistry of National Resonator Instruments* (Fullerton, Calif.: Centerstream, 1994), esp. pp. 114–41.

5. The full history of Latino influence on rock 'n' roll has yet to be told, but see, for a start, David Reyes and Tom Waldman, *Land of a Thousand Dances: Chicano Rock'n'Roll from Southern California* (Albuquerque: University of New Mexico Press, 1998).

6. Richard R. Smith, *Fender: The Sound Heard around the World* (Fullerton, Calif.: Garfish, 1995), pp. 34, 112–13, 141.

7. Tom Wheeler, *American Guitars: An Illustrated History* (New York: HarperCollins, 1990), p. 331.

8. Cordell Jackson, interview by Pete Daniel, Charlie McGovern, and Peter Guralnick, 1992, for "Rock 'n' Soul: Social Crossroads Project," transcript in the Division of Cultural History, National Museum of American History, Washington, D.C., p. 17.

9. Jim Fisch and L. B. Fred, *Epiphone: The House of Stathopoulo* (New York: Omnibus, 1996), pp. 147–48, 150–51; Smith, *Fender,* pp. 22, 46; Wheeler, *American Guitars,* pp. 310, 331.

10. Les Paul, interview by Charlie McGovern, National Museum of American History, at "Electrified, Amplified, and Deified" Symposium, October 1996.

11. Nadine Cohodas, *Spinning Blues into Gold: The Chess Brothers and the Legendary Chess Records* (New York: St. Martin's, 2000), pp. 52–53. Jas Obrecht, *Rollin' and Tumblin': The Postwar Blues Guitarists* (San Francisco: Miller Freeman, 2000), p. 104, excerpts Muddy's version of the delayed move to electric guitar.

12. Bob Dunn's centrality to the steel and to Milton Brown's music is discussed at

length in Cary Ginell, with Roy Lee Brown, *Milton Brown and the Founding of Western Swing* (Urbana: University of Illinois Press, 1994).

13. Mary Alice Shaughnessy, *Les Paul: American Original* (New York: William Morrow, 1993), pp. 35–63. Paul interview by McGovern.

14. See Scotty Moore, with James Dickerson, *That's Alright Elvis: The Untold Story of Elvis's First Guitarist and Manager* (New York: Schirmer, 1997), pp. 47, 51, 100; Peter Guralnick, *Last Train to Memphis: The Rise of Elvis Presley* (Boston: Little, Brown, 1996); Scotty Moore, interview by Pete Daniel and Charlie McGovern, August 1992, for "Rock 'n' Soul: Social Crossroads Project," p. 3.

15. Calvin Newborn, interview by Pete Daniel, Charlie McGovern, and David Less, August 1992, for "Rock 'n' Soul: Social Crossroads Project," p. 49.

Further Reading

King, B. B., with David Ritz. *The Blues All around Me: The Autobiography of B. B. King.* New York: Avon, 1996.

Perkins, Carl, with David McGee. *Go, Cat, Go!: The Life and Times of Carl Perkins, The King of Rockabilly.* New York: Hyperion, 1996.

Reyes, David, and Tom Waldman. *Land of a Thousand Dances: Chicano Rock'n'Roll from Southern California.* Albuquerque: University of New Mexico Press, 1998.

2 inventing the electric guitar

André Millard

Who invented the electric guitar? Most school children know that Alexander Graham Bell invented the telephone and that Thomas Edison devised electric lighting, but the origins of the most important musical instrument of the twentieth century are not as well known. The question is important and much more than academic; historians and collectors are always interested in who was first as a matter of record, but beyond that there are important commercial considerations at stake. Manufacturing electric guitars is a competitive, high-tech business in which history and tradition still play a significant role. The honor of being the father of the electric guitar is an important marketing tool and is contested by long-established concerns such as Fender, Gibson, and Rickenbacker.

The importance of being first to have an idea is based on a legal system that protects intellectual property. U.S. patent laws give credit and protection to the first person to file a claim on a new device, process, or material. In the business of electric guitar manufacturing, the distinction of being the first in the field implies the high ground of technological leadership. Any claim to be the originator of this valuable technology will not go unchallenged; the Rickenbacker company took offense at television advertisements that credited Les Paul as the inventor of the electric guitar. It had its founder's claim to protect, as does the Fender company.

We usually think of invention as a single event in which an amazing new discovery is made. This is a throwback to a nineteenth-century view of inventing: one heroic individual overcomes numerous obstacles to reach a goal that many think impossible. The nineteenth century was, after all, the age of Bell and Edison, of King Gillette and Samuel Colt—heroic individuals who founded their own companies, imprinted their names on their products, and often wrote their own histories. These legendary figures stood in the pantheon of American invention, or to put it in the vernacular of the times, Yankee ingenuity—the secret ingredient that made the United States the fastest growing economy in the world.

This image of the heroic inventor was formed at a time when a handful of individuals garnered the public's attention and most of the credit for the new technology that transformed American life. In newspaper stories, magazine articles, paperback books, and movies the typical inventor was self-taught, worked alone and often in poverty, and had to fight failure and rejection before turning a moment of inspiration into a valuable discovery. Inventing embodied some central values of the American republic: ingenuity and self-reliance, individuality and hard work. Edison's famous pronouncement about invention being 1 percent inspiration and 99 percent perspiration might have completely missed the point about the inventive process, but it touched a chord in nineteenth-century America, where hard work (and a little bit of luck) was believed to be the key to success. Indeed Luck and Pluck was the motto of the widely read Horatio Alger novels, which described the American dream of advancement through talent and effort. Nobody illustrated the workings of this dream better than the great inventors, whose life stories often depicted penniless farm boys or immigrants beating the odds to become millionaires. In the land of opportunity the famous inventors were heroes.

Inventing in the good old days of the nineteenth century was not considered much of a scientific pursuit. For one thing there were very few people at that time that we would consider scientists, and for another many of the heroic inventors deliberately cultivated images that downplayed any connection with organized scientific knowledge. Edison, for example, promoted himself as a self-taught tinkerer who had little public sympathy with higher education and the scientific elite. Although he had one of the best-equipped research laboratories in the world, his admirers thought of him as the man who slept on a bench in a plain workshop. In the public mind the cradles of invention were dusty attics or cluttered workshops, where the work was done by dogged, disheveled inventors rather than clever people with university degrees and white coats.

Although the practice of invention changed radically in the twentieth century, when organized research departments replaced the lone worker, the image of the heroic inventor did not die that easily. The important technology came out of company laboratories and was the work of company men—quietly efficient, university-trained scientists—but there were also a few inventors who fit the mold of the nineteenth-century tinkerer and great American hero. And nowhere is this more evident than in the history of the electric guitar.

The "fathers" of the electric guitar have hardly a university degree among them. Leo Fender took a few courses in accounting at a junior college, Adolph Rickenbacker was a machinist, and Les Paul gave up school to play on a radio show. Most of the people involved in developing the electric guitar came from rural or immigrant backgrounds. Adolph Rickenbacker described his collaborator George Beauchamp as "a young Texas boy [who] got too fat to pick cotton."[1] These inventors had few resources but boundless imagination and ambition. This is the stuff of heroic inventing.

Les Paul has been credited with inventing the electric guitar in books, videos, and television commercials. The Rock and Roll Hall of Fame in Cleveland boldly states that he invented the solidbody electric guitar in 1941. The November 1999 issue of *Guitar* magazine also had no hesitation in crediting Les Paul with this invention. Often described as "a living legend," he is acknowledged, too, for perfecting the technique of multitrack tape recording and overdubbing. His accomplishments have overflowed from technology into culture. In 1988 Cinemax produced a film about his life entitled *Les Paul: He Changed the Music*. As a guitar player of rare talent and amazing longevity, he is an important figure in the world of popular music. But it is probably his perceived role in the creation of the electric guitar which gives his name such weight. The Gibson guitar company enjoys a long and profitable relationship with him, having manufactured a series of classic instruments bearing his name. In their advertisements the company stresses the ties between their guitars and the great inventor.

If inventing the electric guitar was important only in the marketplace for amplified instruments, it would be worth arguing about, but there are other consequences to consider. Technology is sometimes seen as a critical factor in cultural as well as economic development, and because the electric guitar played a part in the creation of rock 'n' roll, the most widely disseminated music of the twentieth century, some observers have discovered a strong relationship between electric guitar technology and changing American culture: "The day an electric pickup was placed on top of an acoustic guitar," claimed one advertisement, "a musical revolution took place."

Fender issued a series of advertisements in the 1970s identifying Leo Fender, the founder of the company, with the statement "This man started a revolution." *Revolutionary* is the adjective commonly used to describe the changes wrought by rock 'n' roll in the 1950s, and the electric guitar, by association, is sometimes credited with producing the sound that "changed the world." This might be an overstatement by an enthusiastic advertising writer, but it does give an idea of the importance of the cultural impact of electric guitar technology. If the inventor and businessman Leo Fender could be credited with starting a "musical revolution," then his guitar did a lot more than produce louder sounds. Clearly the inventor of this instrument can claim a place not only in the history of technology but also in the popular culture of the twentieth century.

fathers of the electric guitar

The search for the inventor of the electric guitar must begin in the late nineteenth century with the first patents for amplifying guitar sounds. One of the pioneers was George Breed, a naval officer who filed a patent for "a method of and apparatus for producing musical sounds by electricity" in 1890. No doubt there were other would-be inventors working in the same field.

The first man who claimed to be the inventor of the instrument was Adolph Rickenbacker, whose business card identified him as "Father of the Electric Guitar." He was an immigrant to the United States, like many of America's most famous inventors. He arrived in 1886, at a time when other immigrants, such as Emile Berliner and Alexander Graham Bell, had made valuable innovations in amplifying and transmitting sound. Adolph Rickenbacker established a thriving tool and die business, and one of his best customers was the National String Instrument Corporation, a manufacturer of metal-body guitars with built-in "resonators" to increase their volume. National had been founded by John Dopyera, an immigrant from Czechoslovakia who had designed the resonating guitar—a first step in increasing the volume of the instrument. The general manager of the company was George Beauchamp, a musician and tinkerer who came from Texas. Rickenbacker joined with Beauchamp and another employee of National, Paul Barth, to produce in 1931 the Frying Pan Hawaiian lap steel guitar, which had an electromagnetic pickup. This instrument now sits in a case in a museum, looking every inch the historic artifact but not much like a guitar (see plate 2). With a long neck attached to a small round body, it does resemble a frying pan, but can this be the first electric guitar?

To decide who invented the electric guitar, one first has to define it: is it a solid-body instrument intended only to amplify the sounds of the strings, or can it be

Adolph Rickenbacker with his baby—the prototype Frying Pan guitar. Note the large magnet above the strings. Courtesy of Rickenbacker International Corporation.

any hollowbody acoustic guitar that has some means of amplifying the sound? If one chooses the latter, the identity of the inventor is submerged in a pool of numerous individuals and companies who were experimenting with increasing the volume of acoustic guitars in the 1930s.

The leading musical instrument makers had their own research and development operations, in common with every other large corporation at that time. Invention had been incorporated into business affairs, and large companies had designated employees whose duties were to improve existing products and devise new technology. Gibson hired musician and composer Lloyd Loar in 1919 to investigate and improve all aspects of their instruments' performance. He was also expected to train employees and the company's agents, to give demonstrations, and to write articles for trade publications. Guitars were only a small part of the Gibson product line, and Loar also worked on mandolins, violins, and banjos. During the 1920s several other Gibson employees, including Thaddeus McHugh (truss rod) and Guy Hart, made patentable improvements to the company's instruments. Loar is remembered for his early work in applying electromagnetic pickups to acoustic instruments. In 1924 he demonstrated an electrified double bass. As a major guitar manufacturer and a leading force in the industry in the 1930s, Gibson set the direction that electrification would take. After Loar left Gibson, his place was taken by Walter Fuller, who worked on Gibson's line of electrified Hawaiian and Spanish guitars and designed its electric pickups.

These engineers were not the only people working on increasing the volume of guitars. Musicians in big bands and blues groups were also experimenting with adding pickups and microphones to acoustic guitars. Whether it was in the laboratories of large companies or individual inventors working in their garages, the technological problem they all faced was a simple one: how to increase the volume of the acoustic guitar. In the 1930s popular music in America was the sound of the big band, a jazz-influenced dance music with a propulsive rhythm, which was appropriately known as "swing." Dancing was an important part of the social life of many Americans—young and old—and the big bands provided the music. Each band was divided into sections: wind instruments, strings, and rhythm. The acoustic guitar was part of the rhythm section, and as bands got larger and the music grew louder, the guitar was overwhelmed by the volume of sound coming from trumpets, saxophones, trombones, and clarinets. In smaller groups of musicians (such as Django Reinhardt's Hot Club de France) the acoustic guitar had held its ground, but in the big swing bands the sound of the guitar could hardly be heard. (Many bands used a banjo rather than a guitar because it was louder.) The guitarist was relegated to background rhythm, playing chords behind the leads of saxophones or trombones. In the words of Charlie Christian, he was "a robot plunking on the gadget to keep the rhythm."[2]

The low volume of the instrument made it harder to reach the big groups of dancers who were the musicians' bread and butter. Its muted tones meant that its player was not as evident or as valuable as the man blowing the horn or beating the drums. Increasing the sound of the guitar meant reaching more paying customers and increasing the visibility and worth of its player.

The problem of amplifying sound had already been faced by the large telephone and phonograph companies; and Edison, Bell, and Berliner had pointed to the solution. The Bell Laboratories' system of electronic amplification using vacuum tubes was adopted by the motion picture producers, the sound recording companies, and the radio industry in the late 1920s. Transducers, which change sound into electric currents, were employed in telephones and microphones. Loudspeakers, which reverse the process, were used in radios and record players. The phonographs and gramophones that stood in the living rooms of millions of Americans had a pickup that translated the movement of the stylus into sound. The radio set had amplifiers and loudspeakers that amplified the sound of voices and music broadcast miles away. As George Beauchamp theorized, "if you can amplify radio waves, why not amplify vibration waves."[3]

For amateur inventors all the technology required to increase the guitar's volume was readily at hand. They were not dealing with revolutionary new concepts but with equipment they had been playing with for years: microphones, record players, radios, and vacuum tubes. The technological pieces were all around them, waiting to be put into place. Even musicians were taking microphones apart and attaching them to guitars. Others were taking the magnets and coils from telephone receivers to make pickups. Amplifiers were being fashioned out of parts of radio sets and record players. Les Paul was not the only experimenter who took the pickup of an electrified phonograph and jammed it into the strings of a guitar to amplify the sound of the strings. Beauchamp got hold of a Brunswick phonograph, took the pickup out, extended the wires, and mounted it on a 2-by-4-inch block of wood with a single string.

The basic idea of a guitar pickup is to turn the vibration of the strings into electric currents. It is a magnetic device activated by the sound waves coming from a guitar string. As the string vibrates (when it is plucked), it moves over the magnetic field of the poles of a magnet and generates minute electrical currents that can be amplified and translated back into sound waves. Inventors investigated the arrangement of magnets and coils to achieve the most efficient transfer

Mission Beach dance hall in the big band era. This was how people used to have fun before the electric guitar took over popular entertainment. You could actually hear yourself flirting with your partner. Courtesy of Library of Congress.

of motion into electricity. In the Gibson electrified guitar the bar pickup had two long magnets attached underneath the soundboard by three screws. Beauchamp's design for the Frying Pan guitar was a pair of large horseshoe magnets that enclosed a pickup coil and strings. Inventors put magnetic pickups above, below, and around the strings of guitars.

Lloyd Loar apparently left Gibson because the company did not see the potential of electrification. He built his own electric guitar, called the Vivi-Tone, in which the pickup was inserted in a drawer fitted beneath the sound board of an acoustic guitar. In a patent filed in 1934 Loar specified that his guitar could be used as a musical instrument or as "a generator" of electric currents—a critical development that anticipated the time when the guitar would be the key element in a technological system that produced and controlled sound.

The other vital innovation in the invention of the electric guitar, and the final break with the acoustic past, was the solid body. The electrified acoustic guitars of the 1930s and 1940s filled the need for more volume, but their performance was marred by unwanted resonance of the acoustic sound box at high volume. With both the wood and the strings of the instrument vibrating, the amplified acoustic produced a loud and annoying feedback hum. The electrified acoustic guitar was a compromise between two completely different amplification systems: one acoustic and one electric. The important step forward was the abandonment of the acoustic system and the adoption of a solid, immovable base for the electric pickup. In the 1930s several individuals experimented with stringed instruments made from solid pieces of wood, taking the same route as George Beauchamp in his experiments. Les Paul strung a guitar string across a 4-by-4-inch piece of solid pine and placed a homemade pickup under the string. When the string was plucked, the note sounded on and on with hardly any reduction in sustain. Paul called this device the Log. When he attached two halves of an Epiphone acoustic guitar to each side of the Log, he had something that looked like a guitar. Paul found out that the look of the new instrument played an important part in its acceptance by businessmen and his audience. He tells the story of playing it without cosmetic side pieces and being rejected by the audience, but with the side pieces attached the same group of people (on a different night) warmed to the sound of the electrified Log (see plate 3).

The Log was an experimental device and never went into production. In the 1930s the only solidbody electric guitars made in any numbers were the steel guitars made by Rickenbacker and his associates. At this time the steel or Hawaiian guitar was as popular as the standard or Spanish guitar. It was attached to a stand or placed in the lap of the player, who moved a slide over the metal strings to achieve a drifting, slurring sound. In 1931 Beauchamp, Barth, and Rickenbacker set up the Ro-Pat-In Company to make these instruments, and in 1934 the name was changed to the Electro String Instrument Corporation. The company (later renamed after Rickenbacker) manufactured several models of electric steel guitars in the 1930s and 1940s, all based on its original Frying Pan design. The company shared production facilities with Rickenbacker's tool and die business in a factory on South Western Avenue in Los Angeles. They produced hundreds of thousands of solidbody guitars made from metal and from Bakelite, an early form of hard plastic. From this successful base in manufacturing Hawaiian guitars the company moved into producing solidbody Spanish style guitars. The Electro String Vibrola guitar was introduced in 1937. The first commercial electric guitars were sold as sets: for $125 you could buy the Electro String Spanish guitar and an amplifier, or for $152 you could get a Gibson ES-150 electric Spanish guitar and amplifier.

The attention of the manufacturers was as much on the electrical circuits of the pickup and the amplifier as on the construction of the instrument. For this reason many of the inventors of the electric guitar were entrepreneurs in the burgeoning field of radio. Leo Fender, for example, was a radio enthusiast who established a business building amplifiers and public address systems in the 1930s to rent to dances, sporting events, and political rallies. He opened Fender's Radio Service in 1938, which also repaired amplifiers and phonographs and sold records.

Another one of the radio entrepreneurs was Francis Cary Hall, who moved from Iowa to California and established a battery manufacturing business. He branched out into radio, founding Hall's Radio Service repair shop and then the Radio and Television Equipment Company (called Radio-Tel for short) to sell parts wholesale. As the distributor for Leo Fender's products, Hall played an important role in the birth of the solidbody electric guitar as a financier and business strategist.

In the 1940s Leo Fender began to collaborate with a fellow would-be inventor, Clayton "Doc" Kauffman, yet another midwesterner who had moved to California in the 1920s. Kauffman was also a musician who played the violin and lap steel guitar. In a small workshop behind his house he worked on various moneymaking inventions: dishwashers, automatic hot water bottles, and a floating seat for tractors. He devised a vibrato unit to change the pitch of the guitar and then patented it. He knew Beauchamp, Barth, and Rickenbacker at the Electro String company because they bought the rights to manufacture this invention, the Vibrola tailpiece. He also produced a version of their Bakelite Spanish guitar with a vibrato unit attached to it. Doc was an ideal partner for Leo Fender, and together

Leo Fender, American inventor, enjoying his work. Photo by Robert Perine, 1966.

they worked on a variety of innovations. He said later that the two men "had more fun than money," but one of their devices, an automatic record changer, brought in enough cash to finance their own manufacturing company.[4] The idea was to market the innovations springing from Fender's fertile workshop: amplifiers, steel guitars, and a pickup for an electric guitar. With the wartime restrictions on raw materials lifted, they custom built amplifiers and steel guitars for a growing number of dealers who drove from Los Angeles to the company's plant in Fullerton.

Although not much of a musician himself, Fender liked music, especially the Hawaiian music of the 1920s. One of his few pastimes outside inventing was hanging around country and western clubs. The sound of the steel guitar had become popular in country music in the 1940s and was especially noticeable in the fusion of country and big band jazz called western swing. The most accomplished bands—Bob Wills and the Texas Playboys or Spade Cooley and his Orchestra—had national reputations and Top Ten hits. The sound of western swing leaned heavily on the fiddle and the electric steel guitar and made stars of the men who played them. Even the upbeat country songs that crossed over into the pop charts in the late 1940s would have a solo from one of these instruments. The steel guitar was by far the most popular electrified instrument, selling in greater numbers than other electrified guitars, banjos, mandolins, or violins. It was the growing demand for steel guitars in California that attracted entrepreneurs like Leo Fender and F. C. Hall.

Although the basic idea of the steel guitar went all the way back to the Hawaiian guitars of the 1920s, there was plenty of room for improvement. They were traditionally played in one open tuning, and this limited musicians. Manufacturers devised steel guitars with foot pedals to change the tuning and also constructed guitars with two necks, each tuned differently. The pedal steel guitar used a foot pedal to raise and lower the pitch of the strings. In 1939 Gibson introduced the Electraharp, which had eight strings and six pedals. By 1946 Fender was manufacturing his single-neck Hawaiian guitars, and the following year he introduced his double-neck Dual 8 Professional model.

Soon Fender made enough sales to sign a marketing deal with F. C. Hall's Radio-Tel Company in 1946 and start manufacturing a line of steel guitars and amplifiers. He found that before long the market for steel guitars was saturated and listened to one of his salesmen, Charlie Hayes, who suggested they produce an inexpensive electrified Spanish guitar. Fender liked the idea of a standard guitar that sounded a little like a steel, with its trademark metallic tone and high sustain. The few models that were on the market, including the Rickenbacker in-

Leo Fender, the full-time tool aficionado, with a punch press at his factory. How can work be this much fun? Courtesy of Richard R. Smith.

struments, did not impress him as good enough for professional musicians. He thought he could do better.

In this view he was not alone, for at least one country and western musician had the idea of improving the electrified Spanish guitars in use. Merle Travis was one of the great guitarists in the country world; his playing style of finger picking was copied by many and is known today by his name. In 1947 he approached machinist Paul Bigsby, who had designed his own pedal steel guitar and had custom built steel guitars for other players (including Speedy West), and asked him if he could build a solidbody electrified Spanish guitar. Bigsby replied that he could make anything.

Fender customer at work. Noel Boggs plays a Fender dual-neck steel guitar in 1946 in the obligatory cowboy shirt. Courtesy of Richard R. Smith.

Aug. 10, 1937. G. D. BEAUCHAMP 2,089,171

ELECTRICAL STRINGED MUSICAL INSTRUMENT

Filed June 2, 1934 3 Sheets-Sheet 1

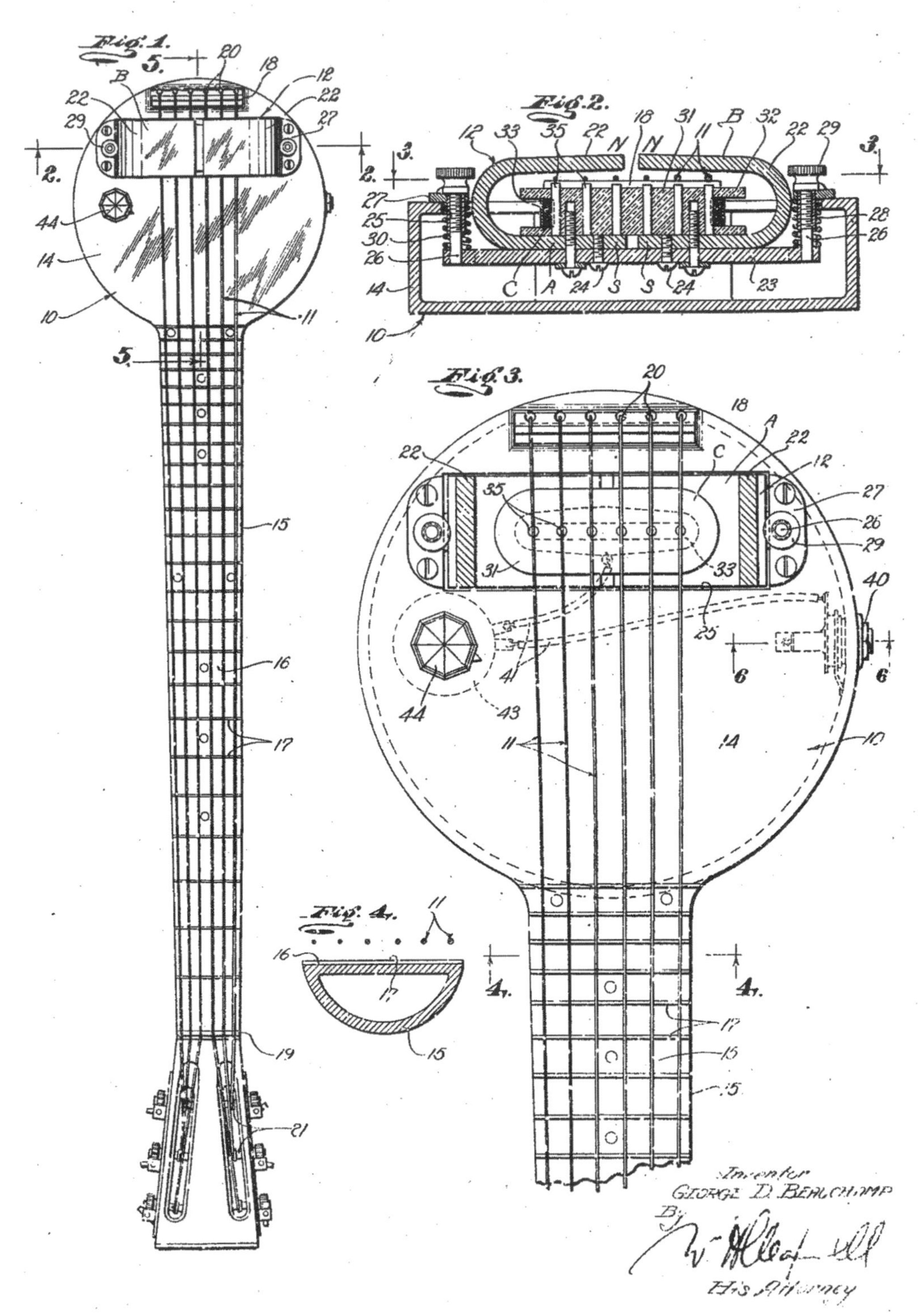

Paul Bigsby represented a long and very important tradition in American inventing. A skilled machinist who could turn his hand to almost everything, he could take the sketched ideas of others and make them into machines. The great nineteenth-century inventors invariably worked with machinists, whose job was to transform rough drawings into working models. The sketch that Merle Travis handed over to Bigsby was of a Spanish guitar with a curved body and an electrical pickup. The tuning screws were arranged along the top edge of the head, an arrangement that had been used in some European instruments but was completely new for American guitars. The guitar that Bigsby made for Travis sounded good and looked impressive. Travis played it in clubs in southern California, and soon other musicians were asking Bigsby to make them copies. It is said that by 1949 several had been made and sold for five hundred dollars apiece.

At this point Leo Fender entered the picture. He saw Travis playing the guitar and lived close enough to Bigsby's shop to come over and see the instrument on the bench. Legend has it that Fender asked to "borrow" the guitar and kept it for some time. The resemblance between Bigsby's guitar and the Fender Telecaster and Stratocaster is strong enough to lead some observers to conclude that Fender did more than borrow the guitar—he stole the idea from Bigsby. Ted McCarty of Gibson thought that Fender "copied" Bigsby's instrument.[5] There can be no doubt that Fender was working on the solidbody prototype, called the Esquire, at this time and that it bore more than a passing likeness to the Bigsby guitar.

No story of invention would be complete without somebody stealing somebody else's ideas or taking the credit for another's work. This too is part of the myth of the heroic inventor. Edison might have been the world's greatest inventor, but he was not above appropriating the ideas of others. Les Paul claims that Fender came to his "backyard" around 1947 and asked if the two of them could go into business together.

Nobody can tell how much of the Fender solidbody guitar (now called the Telecaster) was "borrowed" from the Bigsby guitar, but comparing the two instruments and the motivation of the two inventors is very helpful. Bigsby was a craftsman, a man who made beautiful things one at a time to order; Fender (by this time) was a manufacturer and entrepreneur, a man who had built a factory to mass produce guitars. The Telecaster might have had a resemblance to the Bigsby guitar, but its inner structure and its purpose were completely different. Although Fender was making electric guitars for the professional musician (and not at this time for the general public), he wanted a design that

George Beauchamp's 1934 patent for the Frying Pan guitar—an artifact from the early history of the invention of the electric guitar. Courtesy of Rickenbacker International Corporation and U.S. Patent and Trademark Office.

was easy to make and maintain. Thus his instruments had bolted-on necks that could be easily attached in the manufacturing process and just as easily removed for repair or replacement. This is the important difference between the two. Significantly, Bigsby soon got tired of being asked to make electric guitars and gave it up.

While there is general agreement that the Telecaster was the first solidbody electric guitar to go into mass production, that is probably the only thing the inventors and their chroniclers can agree on. Did Fender steal Bigsby's ideas? Did Beauchamp steal Dopyera's ideas while they were at National, or was it vice versa? Did Rickenbacker take the credit for the work of Beauchamp and Barth on the Frying Pan? It was Beauchamp who was awarded the patent on the Frying Pan guitar, which he first filed in 1932 and then resubmitted in 1934. Accounts from fellow researchers and family members indicate that he did the lion's share of the development of this instrument. Why, therefore, isn't his name connected with the first electrified guitar or associated with any present-day company? Money and marketing might be the answer. Adolph Rickenbacker was a major investor in the Electro String company and he was in charge of manufacture. His name had a much higher recognition factor than Beauchamp's because Adolph's relative Eddie Rickenbacker was a famous World War I fighter ace. (Adolph changed the German spelling of his name, Rickenbacher, to Rickenbacker sometime in the 1940s, perhaps because of the growing anti-German sentiments in the United States.)

In addition to the cast of inventors discussed in this essay, there could well have been other important figures who have been lost to historians of the electric guitar. There were several inventors and musicians working on electrifying guitars in the 1930s. O. W. Appleton is alleged to have developed a solidbody instrument with a Gibson neck around 1941 but (like Les Paul) failed to sell it to Gibson. Even more intriguing, the guitar collector George Gruhn has found an advertisement for a Slingerland solidbody guitar in production before the Telecaster! Is this the missing link in the evolution of the solidbody electric guitar?

contexts of innovation

Much of the work that led to the electric guitar was collaborative: two inventors working together, like Beauchamp and Barth or Fender and Kauffman, or collaboration between musician and craftsman such as Travis and Bigsby. Most common was the cooperation of famous musician and instrument manufacturer, such as Les Paul and Gibson. In cases like these the question is not who was the inventor, but who should be given the credit out of the many different people who contributed to the invention?

By the 1950s research and development (R&D) was an important part of American business strategies. Companies that had once employed a solitary employee to work on new technology (among many other tasks) were now forming R&D departments of professional engineers and designers as they grew larger and produced a wider range of products. Gibson is a case in point. Walter Fuller took Lloyd Loar's place, and by the 1940s he was chief engineer of a department that hired college- or army-trained engineers. In 1939 the company's advertisements could boast that the Electraharp was the result of five years of development in "the Gibson research laboratories." Ted McCarty joined the company in 1950 and remembered teams of engineers working on new guitars.

The moving force behind the creation of the Gibson Les Paul guitar in the early 1950s was commercial rather than technological; Gibson's engineers did not advance the ideas incorporated in the Les Paul Log and bring them to fruition in the form of a new electric guitar. Les Paul had built and played several Logs during the 1940s but could not persuade any guitar manufacturer to adopt the design. "They laughed at me," he recalled. What brought this guitar into being was not a stroke of brilliance by an inventor nor the enlightened support of an entrepreneur, but the basic economic forces of the marketplace. The growing number of Fender Telecasters in use made the Gibson company think again about the commercial viability of a solidbody guitar. Ted McCarty, the president of Gibson, explained, "Fender had come out with the solidbody, or the plank guitar, as we called it, and the rest of the industry was watching." Gibson was not impressed with it, but it did have a tone that one could not get with acoustic electrics and, most important, it was catching on with country and western players. "We finally came to the conclusion that we had better get on the bandwagon," McCarty said. Gibson's version of this invention is that their solidbody electric was designed in-house and only after its creation did the company approach Les Paul. Gibson knew that he had been making solidbody guitars for years and that "he had sort of championed the idea." In the early 1950s Paul was playing Epiphone guitars and using the company's facilities in New York for his experiments. Gibson brought him the new guitar, complete and ready to play, in an effort to recruit him to promote their products.

This story can be rewritten within the context of the heroic inventor: after suffering rejection and ridicule from the businessmen in the field, the lone inventor spends years in the wilderness before the business establishment finally sees the light. The Gibson company executives are ordered to find "the kid with the broomstick and the pickups on it" (in the words of Les Paul), and thus years after the invention the inventor is finally vindicated. "It proved the point," concluded Les Paul. There are even more alternatives to this version. Other stories about the invention of the Les Paul claim that Gibson wanted to manufacture an electric guitar to his specifications but did not want to put the company name on such an innovative product![6]

One issue in the invention of the Les Paul guitar is how much the instrument was the work of Les Paul, the inventor, and how much was the work of Gibson employees. According to Ted McCarty the only part of the new Gibson solidbody that was Les Paul's idea was the trapeze-style tailpiece with a cylindrical bar, which was ultimately replaced because it proved to be impractical. From the company's standpoint, the instrument was the work of engineers, designers, and guitar makers in Gibson's factory.

Les Paul spent many years working on the electrification of guitars, and he was clearly one of the leading inventors in the field, yet his role in the guitar that bears his name might have been small. Although McCarty's view was that "Les had nothing more than his name on it," the company's public position has always been: "Designed by Les Paul—produced by Gibson." McCarty and Gibson were delighted to obtain the sponsorship of Les Paul—one of the most popular and influential guitarists of his generation—and always valued his creative input.

Whatever Les Paul's role in the invention of the solidbody guitar, he has many of the characteristics of the heroic inventor. When he was a boy, his love of experimenting and his fascination with dismembering radios earned him the nickname of the "Edison of Rocky Shoals." In later life he was called the "Wizard of Waukesha," echoing Edison's title of the "Wizard of Menlo Park." Although the two worked in different eras, both had a broad range of interests and both were driven by a potent mix of insatiable curiosity and ambition. Les Paul, like Thomas Edison, knew the importance of self-promotion and the use of media and history in the marketing of an innovation. Both men have been called "American Originals," not only because of the technology they produced but also because they conformed to those heroic characteristics associated with mythical American inventors of the nineteenth century: hard working, independent, ambitious, determined, ingenious, and self-reliant.

Leo Fender also fits the profile of the heroic inventor. With no formal training in the ways of technology, he came to it by a practical rather than a theoretical route. Like Edison he devoted his life to experimenting, and he quickly built a lab in his first factory and spent most of his time there, delegating management tasks to others. His passion was taking things apart and improving them, "perfecting" them, as Edison would have put it. According to his business associate

Forrest White, Fender could neither play nor tune a guitar. His interest was in the electronics of amplifying the sound and in constantly improving a useful technology. Although both Edison and Fender made a lot of money, their real passion was for invention. Both were happiest at the workbench in grubby clothes—indifferent to the passing of time, the demands of family, and the state of their businesses. As White remembered, "Leo showed little respect for anyone whose presence would interfere with his beloved R&D lab work."[7] Both put their names on their products in the same rolling cursive script. In his single-minded devotion to experimenting, his disregard for officialdom, his lack of formality, and even the silly pranks he played on his employees, Leo Fender bears a remarkable similarity to Edison.

In his book on Fender, Richard Smith concludes that he "personified the American spirit of invention."[8] In the same way the electric guitar represented the American way of invention, springing from the ingenuity and ambition of countless amateur inventors, mechanics, and tinkerers. They did not need elaborate scientific equipment, just whatever was at hand: old radio sets, leftover wood, parts of phonographs, spools of copper wire, and a sewing machine motor to wind the pickup coils. They labored in workshops and kitchens, often at night, experimenting for "days and nights without sleep" until that magic moment of invention occurred. According to Nolan Beauchamp, the son of George Beauchamp, the historic Frying Pan guitar was made on "an old beat-up bench in the back of our garage."[9]

The electric guitar sprang from the restless spirit of Americans. It was one of the by-products of the great move westward during the 1930s and 1940s. After the war millions of war workers stayed in California, and rural folk from the Midwest and the South joined them. Inventors and entrepreneurs also moved to California in search of opportunity. They were going to make something out of all the technology that emerged during the war. A new electronic age was in the making, with television as its vanguard, and it was surely no coincidence that Leo Fender came up with the names Telecaster and Broadcaster for his new guitars. They planned to establish businesses based on exciting new technologies. The cinder-block factories that Fender and others erected in the orange groves of southern California were testament to the rewards of such ambitions.

Sources

All advertising copy is taken from the pages of *Guitar Player* and *Guitar* magazines, unless noted otherwise.

Gruhn, George, guitar dealer. Interview by author, 22 May 1998.

Hall, John C., Chairman and Chief Executive Officer of Rickenbacker International Corporation. Interview by author, 27 October 1998.
McCarty, Ted. Interview at National Museum of American History, 15 November 1996, as part of the "Electrified, Amplified, and Deified" symposium, 1996.
Paul, Les. Interview. *GuitarOne,* November 1999, pp. 74–80, 190.
Shaughnessy, Mary Alice. *Les Paul: American Original.* New York: William Morrow, 1993.
Star Licks Sessions with Les Paul. Produced and directed by Mark Freed. 90 min. Hal Leonard, 1993.

Notes

1. Manuscript written by A. Rickenbacker, in Rickenbacker Company Archives, Los Angeles.

2. This statement was made in an article that originally appeared in the *Chicago News,* 1 December 1939, and was reprinted in *Down Beat* magazine on 10 July 1969. Christian noted in the article that the electric guitar earned him a pay raise: when he signed with the Benny Goodman band, his weekly earnings went from $7.50 to $150.00.

3. Tony Bacon and Paul Day, *The Rickenbacker Book* (San Francisco: Miller Freeman, 1994), p. 9.

4. Clayton "Doc" Kauffman, interview in *Guitar Player,* September 1982, p. 21.

5. His comment about Fender copying Bigsby's guitar was found in *The Gibson,* ed. Steve Berger and Christopher Perrius (Essex, U.K.: International Music Publications, 1996), p. 46.

6. Ralph Denyer, *The Guitar Handbook* (New York: Knopf, 1982), p. 58.

7. Forrest White, *Fender: The Inside Story* (San Francisco: Miller Freeman, 1994), p. 90.

8. Richard Smith, *Fender: The Sound Heard around the World* (Fullerton, Calif.: Garfish, 1995), p. 284.

9. Bacon and Day, *Rickenbacker,* p. 9.

Further Reading

Carter, Walter. *Epiphone: The Complete History.* Milwaukee: Hal Leonard, 1995.
Smith, Richard. *Rickenbacker: The History of Rickenbacker Guitars.* Fullerton, Calif.: Centerstream, 1987.

3 manufacturing

expansion, consolidation, and decline

James P. Kraft

At the end of World War I, the manufacture and sale of guitars and other string and fretted instruments in America were activities of small enterprise. Even the largest firms were dwarfed by the chief corporations in steelmaking, coal mining, and other industrial endeavors. Fretted-instrument companies were typically family owned and operated enterprises employing a few skilled luthiers to rout, sand, and glue materials into finished products. In the 1920s, as demand for their products rose, the companies grew, adopting manufacturing techniques that incorporated rapid assembly, technological innovativeness, and more diverse product lines. The Great Depression and World War II temporarily reversed the expansion, but at midcentury the musical instrument business entered a period of unprecedented growth and prosperity. By the 1970s small business enterprise and old forms of craft organization had given way to multinational corporations and the processes of mass production.

During these years small businesses introduced a rich variety of new products that helped generate and sustain their growth. Innovations in guitar manufacturing were especially impressive and significant. In the 1920s technical changes in the shape and design of guitars made the instruments bigger, louder, and easier to play. As a result consumers came to prefer guitars to other fretted instruments, and companies increasingly specialized in guitar manufacturing. The introduction of electric guitars reinvigorated an industry hit hard by the depression and led to larger investments in new plants and equipment after World War II. The development of solidbody electrics made mass production feasible, and the growth of the industry quickened. This growth and innovativeness continued in the 1960s, when giant corporations came to dominate both the manufacturing and the consumer ends of the industry. But in the early 1970s this pattern changed, because the fretted-instrument business, like other sectors of the American economy, entered a period of recession and contraction.

The evolution of guitar manufacturing takes on added significance in the context of the ongoing debates about the historical significance of small business in American life. Historians of American business have generally used the experiences of large enterprises in basic industries such as iron and textiles to understand and interpret their subject, but new studies argue impressively that small business played the central role in national economic development. They say that the flexibility of small firms helped them adjust to new market conditions and encouraged experimentation and innovation. The products and technologies that emerged from small firms were certainly driving forces in economic life, but new theories about the importance of small business must still be viewed as tentative, because many areas of the so-called putative periphery remain unstudied.[1]

Among those unstudied areas is the history of the musical instrument industry, including the branch that produced guitars and other fretted instruments. The history of guitar manufacturing has been written chiefly by guitar enthusiasts, whose work is typically rich in detail but poor in historical analysis. This chapter tries to place the electric guitar, now an icon of American musical culture, within the growing literature on the nature and history of small business enterprise. It shows not only that small firms were inventive and innovative but that they possessed these traits to a greater degree than the large corporations that came to dominate guitar manufacturing in the 1970s.

small businesses

In the early twentieth century the manufacturing and marketing of musical instruments were diffused, competitive sectors of the American economy. Like such industries as those involved in making and distributing machine tools and household furniture, this one consisted of many small or medium-size firms, about 600 of them, scattered unevenly across the country. Approximately 100 of the firms supplied other manufacturers with such parts and materials as piano strings, violin bows, and drumheads. The companies usually employed both skilled and semiskilled workers, the former to handle technical aspects of the manufacturing process and the latter to carry out routine tasks. The industry had almost 40,000 employees in 1904, the vast majority of them wage-earning males engaged in the production process.

The companies embraced a variety of production strategies. Typically, they produced goods in limited numbers on the basis of advance orders. The largest companies had systematized production processes and as many as 1,000 workers. The smallest were artisanal shops of proprietors and handfuls of employees who worked on custom orders. Regardless of their size, the successful companies had the capacity and flexibility to deal with fluctuating consumer demands and periods of boom and bust. The strategy of making a variety of instruments in differentiated product lines was one key to their success.

Piano and organ manufacturing was the largest branch of the musical instrument industry, consisting in 1904 of nearly 450 firms. Of these, about 250 made pianos and 100 made organs, while another 100 made parts and materials for these instruments. Collectively, the piano and organ makers employed 36,000 people, paid wages and salaries totaling $22 million, and had a total capital value of $68 million. The growing popularity of piano-based "ragtime" music early in the century spurred their sales. By 1910 the piano and organ industry had grown to 500 firms and 41,000 employees and had a capital value of $100 million.

In the second decade of the century, stagnant sales and a wave of mergers reshaped the piano and organ industry. Between 1914 and 1919, the number of employees leveled off, and the number of firms declined by about 25 percent, falling from 500 to 372. This growing concentration and a parallel increase in capitalization seemed impressive to those in the industry, but the firms were small compared to those in such basic industries as coal and steel. In 1916 when the stock of American Piano, the nation's largest piano manufacturer, had a value of $10 million, the U.S. Steel corporation was worth more than $1 billion. In that year only one other piano manufactory had a stock value of more than $5 million, and most manufacturers were worth less than $1 million.

Companies that produced string, wind, and percussive instruments formed the second major branch of the musical instrument industry. These companies, which numbered about 180 in 1904, made everything from accordions and banjos to xylophones and zithers, or parts and materials for those instruments. They employed about 2,400 people, mostly men, and had a total capital value of $3.7 million. These figures changed little over the next decade, but around 1914 they began to rise notably. By 1920 the number of firms and workers stood at 240 and 5,000, respectively, and the total capital value of the firms had more than doubled, to $8 million.

Manufacturers of string, wind, and percussive instruments were concentrated in the Northeast and the Midwest. In 1919 almost 30 percent of them were in New York alone and another 20 percent were in Massachusetts and Pennsylvania combined. The midwestern states of Illinois, Indiana, and Michigan accounted for another 20 percent of the firms and about half of the total workforce. Indiana firms alone employed 1,400 workers, many of them at the C. G. Conn Company in Elkhart, the nation's largest manufacturer of brass instruments. During World War I the proportion of women in the workforce in this branch of industry rose slightly, from 17 percent in 1914 to 21 percent in 1919.

The leading string- and fretted-instrument manufacturers in New York included Epiphone, Gretsch, and Favilla Brothers (later Favilla Guitars), as well as John D'Angelico's small shop, which produced some of the nation's best and most expensive guitars. Elmer Stromberg, another highly respected luthier, worked in Boston, as did craftsmen at the Vega Company, which specialized in budget-priced instruments. C. F. Martin, one of the industry's oldest and most respected manufactories, was in Nazareth, Pennsylvania, where it had moved from New York in 1839.

The leading manufacturers in the Midwest included the Harmony and Regal Musical Instrument companies, which made low-priced instruments in Chicago, and the Gibson company, which produced better and higher-priced ones in nearby Kalamazoo, Michigan. Companies in the Midwest benefited from proximity to Chicago, a major rail center and home of some of the nation's largest distributors of musical instruments, among them Tonk Brothers, Chicago Musical Instruments, Targ and Dinner, and the nation's two biggest mail-order distributors, Sears and Montgomery Ward. During the 1920s these mail-order firms sold thousands of low-priced instruments to customers across the nation. Their catalogs advertised dozens of models of mandolins, banjos, guitars, and other instruments. As early as 1916 the mail-order demand was so strong that Sears purchased the Harmony Company.

Despite differences in size and in the markets in which they competed, firms in the fretted-instrument business had many things in common. They were, as already noted, small businesses. Their production plants typically consisted of a single two- or three-story building and no more than a hundred workers. The production world was thus more personal and relaxed than that of steelmaking or automobile manufacturing. Owners were often skilled craftsmen themselves who worked long hours on the shop floor with their employees. The pace of work was not fast, and production figures were generally unimpressive. Only Regal and Harmony turned out more than 100,000 instruments a year. The Martin company, which made only high-quality instruments, produced about 3,000 a year, while D'Angelico and Stromberg made less than 100 each.

Most of these companies made a wide range of instruments. At a time when notions of popular music varied from town to town, manufacturers found it necessary to vary their product lines. Gibson and Epiphone had in fact begun as mandolin makers. Only after World War I did they branch out into banjos and Hawaiian-style steel guitars. By the eve of the Great Depression, the guitar had clearly become the nation's favorite fretted instrument. In 1929 manufacturers produced about 160,000 guitars compared to 140,000 ukuleles, 80,000 banjos, and 30,000 mandolins. The growing popularity of the guitar was one result of the rise of the recording and radio industries, which helped popularize the music of Jimmie Rodgers, "Blind Lemon" Jefferson, W. C. Handy, and other folk and blues guitarists. The fact that some popular banjo players like Eddie Lang switched to the guitar testified to the changing tastes in popular music. Manufacturers responded to the changes by producing more and more guitars.

The musical instrument industry itself was changing in these years. The experiences of companies like Regal, Harmony, and American Piano demonstrated the effectiveness of streamlined production, diversified product lines, and improved marketing webs. By driving down prices and expanding markets, larger firms put increasing pressures on smaller ones. As a result, some small firms went out of business or merged with larger ones. Between 1919 and 1927, the number of firms engaged in the manufacture of string, wind, and fretted instruments declined by nearly 60 percent, from 240 to 100 firms, while the size of the workforce remained level, at about 4,100. In the same years, the value of goods rose 44 percent, from $12.5 million to nearly $18 million. These patterns reveal, among other things, the growing systematization of production processes and the increasing substitution of capital for labor.

The history of the Gibson company typifies these developments. Founded in 1902 in Kalamazoo, Michigan, with a capitalization of $12,000, Gibson had a solid place in the music instrument business by the 1920s. Though Orville H. Gibson is generally considered the father of the company, he was not among the original investors. He was instead a luthier who held a valuable mandolin patent, which he sold with his name to a group of investors for $2,500. With thirteen employees, the original Gibson Mandolin Guitar Manufacturing Company set up shop in an abandoned bakery in Kalamazoo.

Success came quickly. In the early years the company specialized in mandolins, though it also produced guitars and other stringed instruments, including violins. In 1907, when its capitalization had grown to $40,000, Gibson abandoned the original factory and moved into two larger buildings in Kalamazoo. Innovative from the beginning, the company introduced fretted instruments with adjustable bridges and elevated pick guards in 1909 and 1910. It also created a harp guitar, a cross between a small mandolin and a large harp. When banjos became more popular than mandolins, Gibson adjusted its product line accordingly. Its first banjos appeared in 1918, just after Gibson moved its production facilities once more, this time to a new three-story building. These developments strengthened Gibson's position in the marketplace, and by the end of World War I the company was worth more than $100,000.

Gibson remained flexible and innovative, shifting outputs and using new technologies to meet changing consumer demands. The company was in fact among the most innovative guitar makers in the country. Earlier its technicians had pioneered the development of guitars with metal truss rods that reinforced the necks, thereby making possible the manufacture of slimmer necks and easier-to-play instruments. The fact that the truss rods could be adjusted simplified tuning and bowing and reduced the problem of warping. By the mid 1920s Gibson had a sophisticated line of archtop guitars that produced warmer, more pleasing tones than flat-top instruments. The company's most impressive and expensive archtop at the time was the L-5, priced at $275. With its large body and violin-style F holes, the L-5 was particularly loud and thus especially suitable for orchestras. In addition to this luxury model, Gibson produced several smaller, lower-priced guitars with traditional oval soundholes, which retailed from $35 to $125.

In the Great Depression Gibson's profits declined sharply, as did profits throughout American business and industry. Between 1929 and 1932 an estimated 110,000 businesses failed nationwide, industrial production dropped by 50 percent, and unemployment soared. In some parts of the country between one-quarter and one-third of the workforce was out of work. The depression had a particularly destructive effect on the musical instrument business. Between 1929 and 1933 the number of firms in piano and organ manufacturing declined by a nearly a third, from 210 to 143, and the workforce fell 75 percent, from 15,449 to 4,042. Nearly one-third of the string, wind, and percussive instrument manufacturers went out of business; the total dropped from 106 to 72, and the workforce declined from 4,000 to 2,000.

To cope with the reduced demand reflected in these figures, Gibson cut back production and turned to other products, including wooden toys. Nonetheless, the company remained at the cutting edge of guitar manufacturing. By the late 1920s Gibson had improved its low-priced L-0 model guitars by adding rosewood fingerboards, ebony nuts, and saddles. After the stock market crash, it reached out to consumers by introducing a new "Kalamazoo" line of low-priced guitars, mandolins, and banjos. In 1934, as the economy showed faint signs of recovery, Gibson introduced the most elegant acoustic guitar it had ever made, the Super 400. This maple-bodied beauty, with ornate fingerboard, gold-plated tuning keys, and heavy metal tailpiece, was among the best guitars on the market. But at $400 (which included a leather carrying case), its sales were limited.

Despite the continuing depression, Gibson maintained its solid place in the industry because of the quality as well as the price of its instruments. In the mid 1930s the company was producing a greater variety of instruments than ever. It maintained a strong marketing network with the help of New York–based distributors. Its continued success also helped sustain and improve the technical foundation of the fretted-instrument business. Nonetheless, Gibson remained a small enterprise, employing only about 100 people in a three-story factory.

Other firms in guitar manufacturing were also small. By the time of the depression Chicago's Harmony Company had become the nation's largest manufacturer of stringed and fretted instruments. In the early 1930s Harmony's workforce of about 200 had the capacity to make 500,000 instruments a year, but most of its products were cheaply made and sold to wholesalers for as little as $2.50 each. The Epiphone company in New York, which made better products than Harmony, turned out about 75,000 guitars, banjos, and mandolins annually in the depression years. Epiphone was still a family-owned, single-factory operation, which like Gibson relied on other companies to supply much of its production matériel and to distribute its products.

Other important manufacturers were quite small indeed. By the 1930s the Martin company had been producing fine fretted instruments for a century but still employed only about a hundred workers. The shops of John D'Angelico and Elmer Stromberg were smaller still. Stromberg's manufactory in Boston, where some of the nation's best guitarists bought their instruments, occupied only 250 square feet. Newcomers in the business were invariably small. Kay Musical Instruments in Chicago, for example, a major manufacturer of low-cost instruments, had about one hundred employees in the mid 1930s. West Coast newcomers such as National Dobro and Rickenbacker, both of which played pioneering roles in developing electric guitars, had only a few dozen workers.

electric revolution

In the early 1930s electrification of musical instruments was still in a protean stage, and instrument makers had not yet understood that electrifying guitars would revolutionize their industry. Many American homes, especially those in rural areas, still lacked electricity, and manufacturers of musical instruments generally failed to see the potential in this technology. Their emphasis on acoustic instruments dampened enthusiasm for electric guitars, at least for a while. Established companies naturally feared that new products would jeopardize investments in old technology. As a result it was fledgling firms like National Dobro and Rickenbacker, both in Los Angeles, rather than established companies in the East or the Midwest, that took the lead in commercializing electric guitars. Their first electrics were steel guitars with magnetic pickups. Unfortunately, the first efforts at electrification produced instruments that distorted at high volume and were thus impractical to use in large bands and orchestras.

Despite this problem, growing consumer interest encouraged manufacturers to continue efforts to electrify the guitar. Gibson again led the way by introducing two electric Hawaiian steel guitars, the EH-100 and the EH-150, in 1936. The cobalt-magnet pickups in these guitars, sophisticated by standards of the time, were designed by Walter Fuller. Gibson marketed these instruments with a small fifteen-watt amplifier made by Lyon and Healy of Chicago. In 1936 Gibson also brought out its first electric Spanish guitar, the ES-150. Like the company's earlier L-50 acoustic, this new electric had an arched top of spruce maple, tapered F holes, and a one-piece mahogany neck with a rosewood fingerboard. But unlike its acoustic counterparts, the ES-150's amplified sound was especially suitable for big bands. When Charlie Christian began playing this instrument in the popular Benny Goodman Orchestra in 1939, he not only captured the attention of guitar players but also alerted manufacturers to the possibilities of the new technology.

By the eve of World War II other manufacturers had electric guitars on the market. Epiphone's Electar line of Spanish-style instruments, designed by Herb Sunshine, was among the most impressive of these. They featured improved tone control systems and technically advanced pickups. The Kay Musical Instrument Company, a new Chicago firm that competed with Harmony in the low-cost market, also introduced a line of electric guitars. Though the instruments in this line lacked the craftsmanship of those of Gibson and Epiphone, they sold for as little as $36, amplifier included. The Vega and Harmony companies also brought out low-priced electrics; and Gretsch, which then had production facilities in both Chicago and New York, introduced its first electrics in 1940.

Charlie Christian playing his Gibson ES-150, one of the earliest electric guitars to go into production. From the collection of Rock and Roll Hall of Fame and Museum.

By this time doubts about the technical practicality as well as the musical attractiveness of electric guitars had disappeared. Manufacturers and distributors committed more and more resources to electrics, and soon instruments were available to match every market and musical taste. In 1941, for example, Gibson had three electric models on the market (ES-125, ES-150, and ES-300) at prices ranging from $73.50 to $300.00. Epiphone also had models (Coronet, Century, and Zephyr) priced from $40 to $100. Both of these companies also offered electric mandolins, banjos, and steel guitars, but those instruments competed poorly with the new guitars.

World War II delayed the commercialization of electric guitars, just as it delayed the development of FM radio and television. To support the war effort, the government ordered musical instrument companies to revamp their production processes to make war-related goods. Epiphone, for example, began making aircraft parts. Victor Smith, who designed some of the first electric guitars at National Dobro, recalled the changes caused by the war. "We all got letters from the President telling us to halt production," Smith recollected. "They needed the aluminum, steel, and other materials for the war. We went to Washington, started bidding on war production work, and picked up our blueprints."[2]

After the war ended in 1945, pent-up demand, especially for electric guitars, fueled the most dynamic growth in the history of the industry. By 1947 the number of companies making string, wind, and percussive instruments had reached 170, compared to 100 in 1939, while the number of workers had grown from 4,100 to 5,300. During the same period the value of products nearly tripled, from $11 million to $31 million. Clearly the depression in the industry was over.

Gibson fared particularly well in the postwar environment, partly because it had been absorbed by a larger company with ample capital and farsighted management. In a classic example of backward integration, Chicago Musical Instruments (CMI), one of the nation's largest distributors, had taken over the Kalamazoo firm in 1944. Anticipating the postwar boom, CMI's president, Maurice Berlin, enlarged the Gibson factory by more than fifteen thousand square feet and set in motion plans to improve and expand Gibson's product line.

By midcentury Gibson had several new electric guitars, the most significant being the ES-350 and the ES-175. Introduced in 1947, the ES-350 was Gibson's first electric with a rounded, single-cutaway body, which gave players access to higher frets. Like the earlier ES-150 it featured a single pickup located near the fingerboard and separate control knobs for volume and tone (treble and bass). In 1948 Gibson brought out a two-pickup version of the same instrument. The addition of the second pickup, located near the bridge, expanded the range of sounds.

An extra control knob also enabled players to emphasize either the mellow sounds of the original front pickup or the bright, sharper sounds of the new back pickup.

The less expensive ES-175 had a sharp, Florentine-style cutaway body. It too was finely crafted and was soon chosen by some of the nation's best jazz guitarists. Its design made it relatively easy to produce, and in 1950 the company turned out more than 500 of the model, compared to about 50 Super 400s and 100 L-5s. In 1953 Gibson introduced a new version of the ES-175 featuring separate tone and volume controls.

The rising demand for guitars, especially electrics, encouraged other manufacturers to challenge Gibson, but they were not always successful. The wartime death of Epi Stathopoulo, the founder and driving force behind Epiphone, was followed by the eclipse of that company, a longtime Gibson competitor at the top end of the market. Stathopoulo's two sons, who inherited the company, apparently lacked their father's business acumen, and their rivalries soon saddled Epiphone with debt. When the company moved from New York to Philadelphia, it lost some of its best craftsmen; and in 1957 Gibson purchased what was left of it—the name, the patents, and the machinery—for $20,000. Gibson continued to make guitars under the Epiphone name but used the name for lower-priced instruments.

As fretted instrument sales soared in the 1950s, the Fender Electric Instrument Company, a small firm in southern California, introduced a line of solidbody electric guitars that soon revolutionized the industry. The new guitars incorporated major innovations in production technology. Although others may have invented the solidbody, Leo Fender's was the first that became popular; he was, as one historian has noted, the Henry Ford of electric guitars, the man who ushered in mass production.[3]

Fender was a radio repairman who first gained attention among guitar manufacturers during World War II for his work on amplifiers and the conversion of acoustic instruments into electrics. He became an instrument manufacturer himself in 1945, when he founded Kauffman and Fender Manufacturing in Fullerton, California. The firm produced steel guitars and amplifiers and distributed them through a firm in nearby Santa Anna. In 1946 Fender bought out Kauffman, changed the firm's name to Fender Manufacturing Company, and moved production to two small buildings near the Fullerton railroad station. He soon had a workforce of 15 to 20, and his guitars and amplifiers gained growing respect among musicians, especially those in the increasingly popular genre of country and western music.

Fender's first Spanish-style solidbody electric, which salesman Don Randall dubbed the Esquire, debuted in 1950 at the retail price of $139.95. Because of problems with the neck, critics claimed that the first models were better canoe paddles than guitars, but Fender blunted the criticism by reinforcing the neck with truss rods. In early 1951 he introduced an improved model, the two-pickup Broadcaster, priced at $189.50. The name infringed on the trademark of a new line of drums and banjos produced by Gretsch, so Fender renamed it the Telecaster.

Telecasters may have looked like works of art to musicians, but skilled luthiers cringed at the implications of the technology that made them possible. Unlike the skilled craftsmanship that lay behind the guitars of D'Angelico and Stromberg, the technology Fender used enabled him to mass produce solid bodies. Using heavy punch presses, semiskilled workers stamped out the cured ash bodies, which required little additional shaping or sanding. Workers then bolted a one-piece neck onto the back of the body, covering the four screws with a chrome jack plate. The rest of the hardware, including tone controls and pickups, was then easily and quickly installed, and the result was a sturdy, easy-to-repair instrument that was indeed the Model T of electric guitars.

But it was more than that. Unlike the traditional hollow-bodied electrics, which tended to produce feedback at high volume, Fender's solidbodies produced clean, crisp sounds with the sustaining power of horns and brasses. With "micro-adjustable" bridges, they also stayed in tune well, their slim necks were comfortable to use, and their tone control was impressive. Their solid bodies also made them more durable than other guitars. Soon guitarists in some of the nation's best bands and orchestras were playing them. A few, including country and western stars Spade Cooley and Jimmy Bryant, used them because they were paid to do so, but most did so because they preferred them to other models. Early advertisements testified to the swift acceptance of the Telecasters. "Dealers who saw them at the music show last month," read one 1951 advertisement, "placed orders for many more than we anticipated. There must be a reason!" There were in fact many reasons.[4]

The success of the Telecaster encouraged other innovations at Fender. In 1951 the company introduced a solidbody electric bass guitar with matching amplifier. The Precision Bass was the first electric bass with frets, and its acceptance helped overturn a long musical tradition. Compared to traditional acoustic basses,

Fender's tremolo patent for the Stratocaster. The six-piece bridge is broken down in the lower drawings. Courtesy of U.S. Patent and Trademark Office.

April 10, 1956 C. L. FENDER 2,741,146

TREMOLO DEVICE FOR STRINGED INSTRUMENTS

Filed Aug. 30, 1954

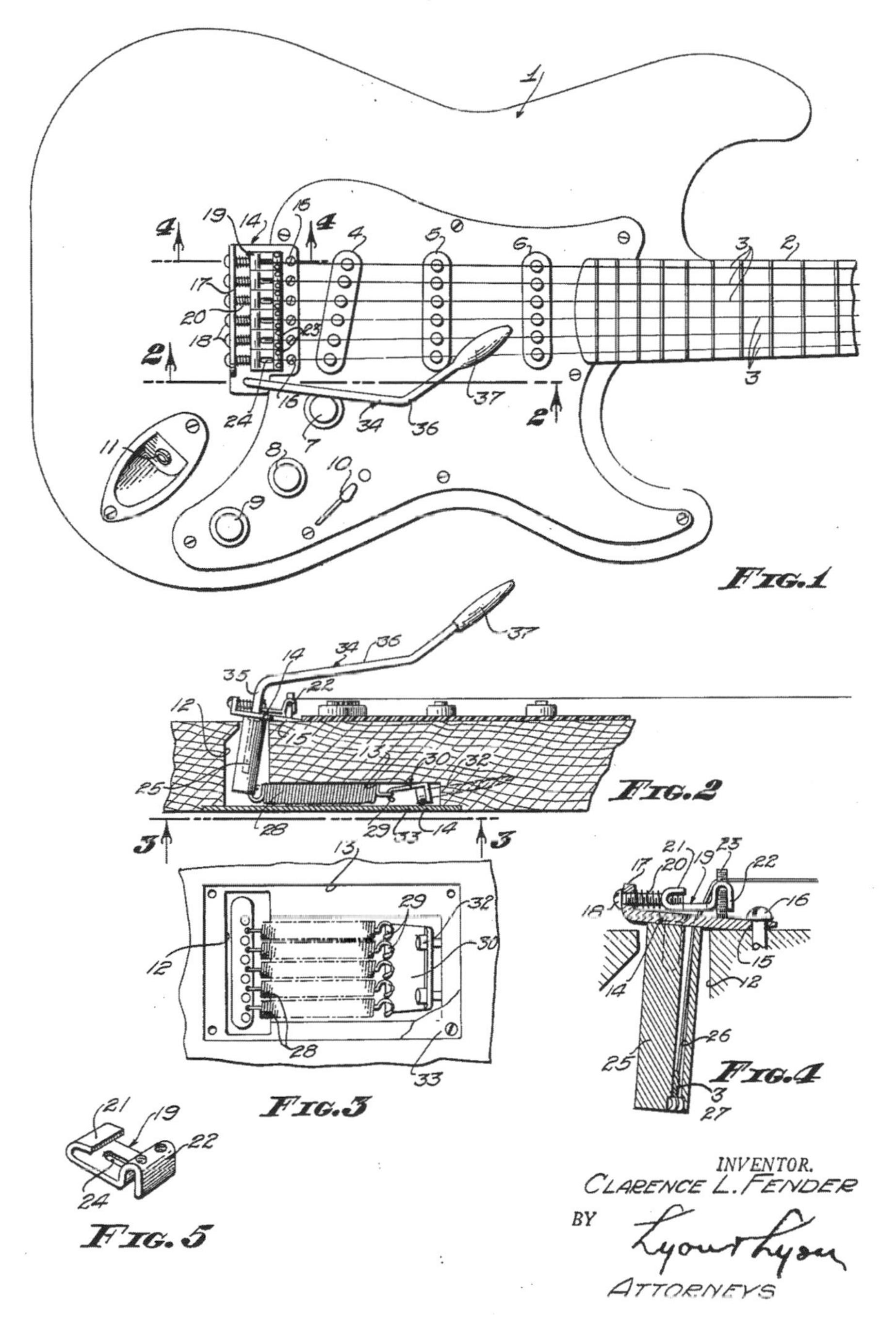

played in an upright position, it was smaller, louder, and easier to play and handle. Priced at $195, it increasingly replaced the "old dog-house" among bassists.

In 1954 Fender introduced still another innovative instrument, the sleekly contoured Stratocaster. It featured three newly designed pickups, which gave it a more biting tone than the Telecaster, and a new six-piece bridge that improved intonation. It also had a tremolo bar, which helped players create wavering, vibrato effects much like those produced by steel guitars. Originally priced at $179.95, the Strat became immensely popular, especially among young musicians looking to generate a style of their own.

The success of these guitars fueled rapid growth at Fender's company. At midcentury, when the company was still making no more than 3,000 instruments a year, Leo Fender expanded his factory space, and he did so again in 1953. In that year he also created a distribution division, Fender Sales, capitalized at $100,000, and a year later he hired Forrest White to manage the company. A former engineer at Goodyear Aircraft in Ohio, White brought a new sense of discipline and efficiency to a production process that desperately needed overhauling. "The place looked like a complete mess," he later recalled of his introduction into the company. "There was absolutely no evidence of work flow. Amplifier and guitar assembly benches were all mixed together—no separation at all. When someone ran out of parts, he got up slowly and walked down to the stockroom in Building #4 and tried to find his own." White also recalled the slow pace of work at Fender. "It was obvious from the way [the employees] were working," he remembered, "that they were not on an incentive plan." Soon after White's arrival, a "Special Bulletin" announced that the company would reduce costs "through work simplification, improved methods and procedures, and elimination of waste." As White gained control over the production process, Fender shed its image as a fledgling company on the periphery of the industry and became a well-managed, vertically integrated business as well as a technologically innovative enterprise.[5]

expansion

Gibson now worked to keep up with Fender. In 1950 Gibson's new president and general manager, Theodore McCarty, decided to develop his own solidbody guitar. The success of Fender's Esquire no doubt encouraged him to do this faster than he otherwise would have done. The resulting Les Paul model, introduced in 1952, was a single-cutaway, gold-colored instrument that sold well despite a retail price of more than $200.

Compared to Fender's solidbodies, the Les Paul was a complex instrument. It had a two-piece, archtop body made of mahogany and maple, a distinctive "trapeze-bridge" tailpiece that helped players "dampen" the strings, and two "soap-bar" pickups with individual tone and volume controls. Although the body was smaller than those of Fender's models, the density of the wood made it heavier and gave it greater sustaining power and a deeper, richer tone. Les Paul's signature on the headstock further distinguished it.

Gibson was soon a major player in the solidbody market. In 1954 Gibson segmented the market with a new series of Les Pauls, priced from $99.50 to $325.00. The all-black Custom model (initially called the Deluxe), was the top of the line. With three electric pickups, gold-plated hardware, ebony fingerboard, and pearl inlays, the Custom was more luxurious than the original Les Paul, which soon came to be called the Standard. At the bottom of the line was the Les Paul Jr., a small instrument with a flat-top body and a single pickup. In 1955 the company brought out a two-pickup version, the Special. A number of new accessories, among them tune-o-matic bridges, "stop" tailpieces, and humbucker pickups, further enhanced the line. Gibson also developed a solidbody bass, but the violin-shaped instrument never sold well and the company dropped it.

While focusing on solidbody guitars, Gibson continued its role in other areas of guitar manufacturing. In 1955 the company introduced a new line of archtop acoustic-electrics that featured smaller, thinner bodies than earlier instruments. This Thinline series included the ES-225T, the ES-350T, and the beautiful Byrdland, which sold for about $500. At the same time Gibson expanded its respected J-series of large-body acoustics.

As Fender and Gibson expanded their operations, other manufacturers found their own ways to capitalize on solidbody designs. Harmony and Kay continued to target beginning players and cost-conscious buyers with new models like the Stratotone. Gretsch introduced a line of solidbodies that included the Duo Jet, the Jet Fire Bird, the Silver Jet, and the Round-up. Rickenbacker brought out one- and two-pickup solidbodies, the Combo 600 and the Combo 800, and offered variations of these instruments in the middle and late 1950s.

In those years new companies entered the solidbody market. One of these, Valco, had been formed in the early 1940s by three former employees of National Dobro. In the mid 1950s Valco marketed three low-priced solidbodies, the best of which, the two-pickup Supro Dual Tone, sold for $135. Another new company, New York–based Guild, also offered its own models. Alfred Dronge had recently made Guild a competitive company by hiring several Epiphone luthiers. In 1956, when Dronge shifted his manufactory to Hoboken, New Jersey, Guild produced several impressive acoustics and acoustic-electrics. Most elegant of them was the large Stuart 550, endorsed by jazz guitarist Johnny Smith.

By the mid 1950s, the fretted-instrument business was larger and more complex than ever. The advent of solidbodies had added a new dynamic to the industry, reorienting it toward larger firms with increased capitalization. More and more of these firms specialized in guitar production, though they continued to produce mandolins, banjos, and ukuleles in reduced quantities. With sales rising, manufacturers faced the future optimistically. None of them, however, foresaw the rapid growth that was about to occur. That growth was a result of the advent of rock 'n' roll, the guitar-based genre that soon changed music culture and with it the musical business.

Changes in the business had mirrored general patterns in the national economy. The period from 1945 to 1970 was an exceptionally prosperous time. Some business leaders had feared a new depression when World War II ended, but the nation's economy expanded steadily for a generation. Between 1945 and 1960 the gross national product doubled, and it nearly doubled again during the 1960s. This sustained prosperity was partly the result of the war itself, which had exacted a heavy toll on all other industrialized economies. While the Britons, the Germans, and the Japanese struggled to rebuild their economies, American manufacturers captured unprecedented shares of the world's industrial markets. In 1970 the nation still produced 42 percent of the world's manufactured goods.

During this prosperity few branches of the economy fared better than string, wind, and percussive instrument manufacturing. Between 1947 and 1969 the value of goods manufactured in this industry rose from $31 million to $172 million. Sales of guitars, especially electric models, soared. By 1964, as Leo Fender said of his own plant, production facilities were "bursting at the seams." That year American manufacturers produced 600,000 guitars, twice as many as at midcentury. Companies like Fender and Gibson, as well as Harmony and Kay, more than doubled the physical size of their operations in these years. Workforces grew too. Between 1954 and 1964 the number of Fender's employees grew from 50 to 400, and the number at Kay and Harmony grew to 500 and 600, respectively.[6]

These increases were not the only measures of growth in the industry. Manufacturers also expanded their distribution systems and gained greater control over the materials they used in production. In the late 1950s, for example, Fender built a large new sales office in Santa Anna, California, and formed a company in Oklahoma to improve customer service and facilitate the distribution of products. A few years later Fender purchased the V. C. Squier Company, which manufactured strings for Fender's instruments. Kay also integrated forward toward consumers and backward toward raw materials. Kay lacked its own distribution company until the 1970s, relying on independent distributors. In 1964, when Kay relocated its production facilities to spacious new quarters near Chicago's O'Hare Airport, the company was using forty-five distributors to get its instruments to retailers.

Manufacturers spurred this growth by introducing new models and accessories. In the late 1950s Gibson brought out a new line of archtop semi-solidbodies, which had the look and feel of thin-line acoustic-electrics but, thanks to a block of wood inside the body, had the solidbody sound as well. The most popular instrument in this line was the ES-335T, a double-cutaway shape with maple body, mahogany neck, and rosewood fingerboard. Several major artists, including Chuck Berry, used it. Other companies followed Gibson's lead. Fender introduced the Jazzmaster in 1957, a solidbody with a distinctive shape, tremolo, and circuit system. Then in the early 1960s the company unveiled the Jaguar and two new basses, the two-pickup Jazz Bass and the six-stringed Bass VI, as well as a new line of acoustics. Meanwhile, Gretsch trumpeted its new Chet Atkins acoustic-electrics and made significant changes in the shape and design of its solidbodies.

Clever advertising promoted these products. By midcentury advertising was an important force in American consumerism. From 1945 to 1950 annual advertising expenditures for American business jumped from $2.9 billion to $5.7 billion, as advertisers encouraged consumers to think, consciously or unconsciously, that they would be getting more than the actual product advertised. Advertisements for Coca-Cola and Pepsi, for example, suggested that soft drinks, basically mixtures of sugar and carbonated water, provided spiritual or physical lift and even a sense of individuality. Other advertisements for everything from shampoos to sports cars had similar messages that played subliminally on the insecurities as well as the hopes of consumers.

Advertisements for guitars associated the instruments with notions of freedom, adventure, and social mobility. Fender advertisements showed people carrying Fenders on airplanes and boats, even while skydiving. "You won't part with yours either," the advertisement explained. Some Fender advertisements suggested that playing the guitar improved one's social life by bringing people together. One of them associated the curves of the guitar with those of a young woman's body, and another showed father and son sharing the day—and a guitar—together.[7]

The effectiveness of these advertisements was enhanced by the advent of television, which not only increased exposure to the advertisements but helped popularize rock 'n' roll as well. In the 1950s televised performances by new stars like

Bikini girl and Fender Mustang guitar. This advertisement is a play on the sensuous curves of the electric guitar, making it clear that this southern Californian beauty is good for posing rather than playing. Photo by Robert Perrine, 1966.

Elvis Presley, Ricky Nelson, and Buddy Holly inspired thousands of people to buy guitars; and in the 1960s new musical technologies, including FM (frequency modulation) broadcasting and long-playing records, as well as such organizational innovations as highly publicized concert tours by popular performers, created a business and cultural climate that benefited guitar makers. By the late 1960s American manufacturers were producing nearly 600,000 guitars a year. But the boom was about to end.[8]

consolidation

The robust profits of guitar manufacturers had attracted attention in boardrooms of large corporations seeking to diversify their investments. Ever since the end of World War II, big businesses in America and elsewhere had pursued strategies of diversification to guard against the threat of falling profitability, and this trend accelerated in the 1960s. When the trend reached guitar manufacturing, it profoundly changed the nature and structure of the industry.

Between 1965 and 1970 large corporations acquired five of the nation's leading guitar manufacturers. Fender went first. In late 1964 Columbia Broadcasting System (CBS), which had extensive holdings in radio, television, and recording, offered Leo Fender a whopping $13 million for his company. Announcing the resulting sale on 6 January 1965, the Fullerton *Daily News Tribune* highlighted Fender's recent successes. The company's products were in such demand, the paper noted, that it was "as much as five months behind on some of its orders."[9]

Other buyouts soon followed. In 1966 Guild became a subsidiary of Avnet, a conglomerate with investments in automobiles, consumer electronics, and other sectors of the economy. In 1967 Seeburg Corporation, a Chicago-based manufacturer of jukeboxes, purchased Kay Musical Instruments; and Baldwin Piano and Organ Company, headquartered in Cincinnati, bought Gretsch. Even Gibson, the giant of guitar manufacturing, changed hands. In 1969 the Ecuador-based ECL corporation, which had primary interests in the beer and cement businesses, purchased Gibson's parent company, Chicago Musical Instruments.

These changes mirrored patterns in the larger economy. Business mergers had been commonplace in America since the nineteenth century, but they became particularly widespread in the late 1960s. In 1968 alone approximately twenty-five hundred mergers occurred, more than a quarter of them representing striking cases of diversification. National and multinational conglomerates became increasingly important components of the economy, and they found small, profitable firms like Fender and Gibson especially attractive.

The buyouts had immediate effects. The conglomerates were more concerned with profits than craftsmanship, and they cared less about technological innovation and quality control than managers like Fender's Forrest White. They were not much interested in the impact of technological change and business consolidation on employees. Their accountants and financial advisers measured success in terms of profits, for the takeover of companies like Fender had been based solely on financial calculation. William Paley of CBS acknowledged this after the purchase of Fender: "We didn't know very much about the manufacturing business, and much more importantly, we didn't care about it. . . . it wasn't our cup of tea."[10]

Corporate buyout was often followed by cost-cutting initiatives. What happened at Fender was typical. In 1966 purchasing agents ordered cheaper coil wiring for Fender pickups. As a result, the insulation on the coils was often damaged during the production process, and the sound produced by the guitars was distorted enough to be detectable. At the same time CBS-Fender began placing larger decals on its guitars and had to enlarge the headstocks of Stratocasters to fit them on. Guitarists thought that destroyed the visual balance of the guitar, and many of them were soon expressing a preference for pre-CBS instruments and accessories.

The restructuring of the industry had major implications for production workers, though the nature and extent of those implications have never been adequately studied. The conglomeration of capital often produced economies of scale, but it also lowered the skill levels necessary to manufacture guitars and thus compromised the bargaining power of workers. Workers in guitar production thus came to share the same sense of powerlessness and frustration that those in other industries faced in the aftermath of similar changes. Many of them faced layoffs or new rigidities in the workplace. Some resigned themselves to the changes, while others resisted as best they could, and a few left the industry altogether.

What happened at CBS-Fender illustrates these patterns. After CBS took control of Fender's Fullerton plant, at least a few of the workers there felt marginalized, even alienated. Forrest White, whose work as general manager of the plant had given him a sense of self-realization and satisfaction, recalled his apprehensions about the new managers. "These guys from the Big Apple," White said later, "were going to show us country bumpkins how the big boys operated. Too bad we had to find out." White's apprehensions were justified. CBS-Fender not only demoted White but cut his salary by a third and ignored his warnings about quality control. After a year White quit the company in a dispute over the design of new amplifiers. "I had too much respect for Leo," he said of the company's founder, "to have any part in building something that was not worthy of having his name associated with it."[11]

Traveling salesmen who worked for Fender also felt the heavy hand of the new management. In the late 1960s CBS executives directed Don Randall, president of the new Musical Instruments Division, to lower the commissions of the salesmen. Randall, who had himself headed Fender Sales before the CBS buyout, feared the effect of the reduction, for sales of Fender instruments had been steady and the sales staff effective. The salesmen had considerable autonomy in their territories and were quite well paid. The 10 percent commissions they earned enabled some of them to make more than the top executives of the company. In fact, the CBS report that recommended the purchase of Fender attributed the success of the company to the system of compensating the salesmen. These men, the report stated, "are probably very well suited for the type of trade they call on and are obviously doing a good job. There has been little turnover owing principally to excellent compensation earned." Randall lowered the commissions, as CBS directed, but the action apparently left him disenchanted, and in 1969 he left CBS-Fender.[12]

decline

In the 1970s conditions in the industry justified cost cutting. After twenty years of steady growth, the industry, like the economy generally, had begun to stall. The revitalization of the economies of other nations, especially those of western Europe and East Asia, was placing new pressure on American manufacturers. The diversification strategies that created large concerns saddled them with high debts and structural rigidities. In 1970 the value of goods shipped by musical instrument manufacturers dropped 8 percent, from $475 million to $437 million. There are no separate figures for guitar manufacturing, but indications are that cutbacks in the production of string, wind, and percussive instruments accounted for about half of the lost value. The value of goods in that branch of the industry fell from $172 million in 1969 to $151 million in 1970. The personnel changes at Fender and other companies were no doubt related to these declines.

The importation of inexpensive Japanese guitars, which began in the late 1960s, was particularly harmful to manufacturers and distributors who concentrated on the low end of the market. Japanese guitars, like Japanese automobiles and machine tools, compared favorably with their American counterparts, and most of the electric guitars imported into the country in these years came from

Japan. "It is no secret that imports, particularly in the guitar field," reported *Down Beat's NAMM Daily* in 1969, "have become a force to be reckoned with in the music market." In that year imports exceeded domestic production for the first time.[13]

In this changed environment the domestic industry collapsed. In 1970 Martin purchased New England–based Vega, which had operated since the turn of the century, and moved its production equipment to Martin's own plant in Pennsylvania. At the same time Favilla Guitars closed down, as did Harmony and Kay. In 1967 Seeburg Corporation sold Kay to Valco, and creditors shut down Valco itself in a few years. Harmony, too, closed a few years later, also the victim of burdensome debt. Charles A. Rubovits, who headed Harmony in the late 1960s, recalled the collapse of the company. "The factory equipment and other assets were sold at auction to satisfy the creditors," he remembered, "and they finally liquidated, paying off ten cents on the dollar." Competition from Japan, combined with poor management, Rubovits believed, explained the collapse. "We had the know-how, but we didn't have the guts," he continued. "We were very successful [in the 1960s], so we didn't want to do more, and as a result the Japanese began to get a bigger and bigger share of the market."[14]

As Rubovits's remarks suggest, the domestic industry had failed to adapt to changing market forces. Even Gibson, historically the most innovative American firm, struggled in the face of the challenge. The most notable of its new guitars in the late 1960s and early 1970s were reissues of the Les Paul, which the company had shortsightedly discontinued in the early 1960s. CBS-Fender was even less imaginative in the new environment. Its most notable new product in these years was a solid state transistor amplifier, which flopped in the marketplace. The new amplifier lacked the warm tone of previous tube-type models, was prone to breakdown, and was difficult to repair. The company's new Coronado series, a line of acoustic-electric guitars, also fared poorly.

Such failures encouraged manufacturers to narrow product lines and relocate production facilities to reduce labor costs. What happened at Gibson was again representative. Just before its purchase by ECL Industries in 1972, Gibson had reduced its budget-priced Epiphone line of instruments from fifty to thirteen models. At the same time the company contracted with a Japanese firm, Matsumoku, to manufacture Epiphone instruments in Japan. In a similar move Gretsch's parent company, Baldwin Piano and Organ, shifted production of Gretsch instruments from New York to Booneville, Arkansas, in the early 1970s and then to Juarez, Mexico, a few years later. CBS waited until the early 1980s to establish Fender Japan, a joint venture with two Japanese firms, Kanda Shokai and Yamano Music.

In this new environment of multinational enterprise and global competition, several small new American firms carved out limited shares of the domestic market. Some of these, like Micro-Frets, lasted only a few years, but others, such as Alembic, Mosrite, and Loprinzi, were more successful. One of the most successful was Ovation Instruments, founded in Connecticut in 1967 by Charles Kaman. Ovation introduced a line of cone-shaped acoustics made from synthetic fibers. These distinctive instruments soon gained widespread respect among musicians, and Ovation grew accordingly. Two other successful new companies, Peavey Electronics and Music Man, began by producing amplifiers alone and then moved into guitar manufacturing. Compared to industry leaders, who had major investments in the status quo, the new firms were more willing to innovate, even to break down established ways of doing business.

During the quarter-century that followed World War II, the musical instrument industry changed profoundly. From 1947 to 1972, the number of domestic firms producing string, wind, and percussive instruments dropped by more than half, yet the value of finished products in the industry quadrupled and the workforce grew by a third, to 7,200 employees. These patterns of industrial concentration and growing employment reflected, among other things, the inventiveness of entrepreneurs in guitar manufacturing. From the 1920s to the 1970s, guitar manufacturing was the most innovative sector of the musical instrument industry. Small firms in this sector adjusted to changing times and new business conditions. They generated the technological innovations that made guitar music more popular and mass production of guitars practical. By the 1960s several of these firms had expanded the scope of their business operations beyond their wildest expectations.

Ironically, the success of these small firms ushered in their demise. In the 1960s large multinational corporations pursuing strategies of product diversification purchased companies like Gibson and Fender. As they did, the pace of technological change seemed to slow, while competition from abroad rose sharply. By the early 1970s, when big businesses controlled the lion's share of the domestic market, guitar sales began to decline sharply. From 1972 to 1989, a time of growing economic problems in the United States, the number of guitars and other fretted instruments sold dropped continually.[15] In the age of conglomerates in the 1980s, for reasons that are very much in need of investigation, the entire musical instrument business hit hard times.

Sources

Carter, Walter. *Epiphone: The Complete History.* Milwaukee: Hal Leonard, 1995.

———. *Epiphone: The Complete History of an American Icon.* Los Angeles: General Publishing Group, 1981.

———. *Gibson Guitars: 100 Years of an American Icon.* Los Angeles: General Publishing Group, 1994.

Gruhn, George, and Walter Carter. *Electric Guitars and Basses.* San Francisco: Miller Freeman, 1994.

Smith, Richard. *Fender: The Sound Heard around the World.* Fullerton, Calif.: Garfish, 1995.

———. *Rickenbacker: The History of Rickenbacker Guitars.* Fullerton, Calif.: Centerstream, 1987.

All statistics, unless otherwise noted, are from the Bureau of the Census:

U.S. Bureau of the Census. *Census of Manufacturers: 1947.* Vol. 1, *General Summary.* Washington, D.C., 1950. P. 799.

———. *Census of Manufacturers: 1972.* Vol. 2, pt. 3, *Industry Statistics.* Washington, D.C., 1976. Pp. 39B-11, 39B-19, 39B-21.

———. *Thirteenth Census of the United States.* Vol. 8, *Manufacturers, 1909.* Washington, D.C., 1913. Pp. 971–73, 984–86.

———. *Fourteenth Census of the United States.* Vol. 10, *Manufacturers Reports for Selected Industries.* Washington, D.C., 1923. Pp. 984–86.

———. *Fifteenth Census of the United States: Manufacturers.* Vol. 2, *Reports by Industries.* Washington, D.C., 1933. Pp. 1327–30.

———. *Sixteenth Census of the United States: Manufacturers, 1939.* Vol. 2, pt. 2, *Reports by Industries, Groups 11–20.* Washington, D.C., 1942. Pp. 578–79.

Wheeler, Tom. *American Guitars: An Illustrated History.* New York: HarperCollins, 1990.

Notes

1. See Philip Scranton, *Endless Novelty: Specialty Production and American Industrialization, 1865–1925* (Princeton, N.J.: Princeton University Press, 1997).

2. Wheeler, *American Guitars,* p. 308.

3. Ibid., p. 67.

4. See advertisement in Smith, *Fender,* p. 94.

5. Forrest White, *Fender: The Inside Story* (San Francisco: Miller Freeman, 1994), pp. 62–65.

6. See Wheeler, *American Guitars,* pp. 234, 240; and Tom Wheeler, "The Legacy of Leo Fender: Part Two," *Guitar Player,* September 1991, pp. 88–92.

7. See Fender advertisements in Smith, *Fender,* pp. 169–70, 177–79, 206, 229.

8. Charles A. Rubovits, Harmony president, interview in Wheeler, *American Guitars,* p. 234.

9. Smith, *Fender,* 249; and *Fullerton (Calif.) Daily News Tribune,* 6 January 1965.

10. Quoted in Smith, *Fender,* p. 252.

11. White, *Fender,* pp. 153, 174.

12. Smith, *Fender,* pp. 253–55.

13. Wheeler, *American Guitars,* p. 234.

14. Ibid., pp. 236–37, 242, 354.

15. Smith, *Fender,* p. 259.

4 solidbody electric guitars

André Millard

Rock 'n' roll was in its infancy when the first Broadcasters and Telecasters rolled off Fender's assembly lines. They were joined by Gibson's Les Paul Goldtops in 1952—years before someone thought of coining a name for the some of the music they were making. Rock 'n' roll had been around as African American slang since the beginning of the century, but its label for youth music originated in 1954, when a disc jockey called Alan Freed decided that the new dance music that his teenage fans were clamoring for needed a more exciting name than the racially loaded "rhythm and blues." It was significant that the name *rock 'n' roll* was made up by a radio personality, not a musician or a record producer or a guitar manufacturer.

In the early 1950s rock 'n' roll was more the sound of the piano and the saxophone than the guitar. The raucous honking of a tenor sax was present in many of the recordings that compete for the honor of being the first rock 'n' roll record. "Rocket 88" by Jackie Brenston was made in 1951 and prominently featured the sound of the electric guitar. The creative force behind the record was Ike Turner, whose innovations in playing the electric guitar have already been noted. His significant contributions to this record are described in chapter 5. Despite its status as a landmark in the evolution of the electric guitar sound, "Rocket 88" is still dominated by the saxophone. To a casual listener, the most memorable parts of this record are the vocals by saxophonist Jackie Brenston and an inspiring solo from tenor sax player Raymond Hill.

The R&B records that were enthralling white teenagers depended on the full-bodied roar of the saxophone to thicken the sound and increase the excitement. The lead work on the first crossover hits was usually done by a pianist or a saxophone player. Many of the players in the R&B or jump blues combos came from the big bands, which were no longer as popular as they had been in the 1940s, yet the practice of having a wind or brass instrument in the foreground did not end with the decline of the big band sound. Louis Jordan and his Tympany Five were among the first jump blues acts to attract both black and white audiences in the early 1940s. Jordan sang and played his alto sax at the front of the band, but the guitar player stayed in the back with the rhythm section.

Many of the musicians who were later credited as rock 'n' roll pioneers worked in the country idiom and had no idea that they were producing new music—it was just the old music played a bit faster and louder. We have a term to describe the proto–rock 'n' roll country, *rockabilly,* but where this started and honky-tonk ended is anybody's guess. The hit record "Rock around the Clock," which started the rock craze in 1955, was performed by Bill Haley and his Comets, once a western swing band but now full-fledged rockers with a lineup that included saxophone, steel guitar, stand-up bass, and accordion in addition to Danny Cedrone on lead electric guitar.

While Nashville was making traditional country music, reverent and melodic, Bakersfield and other industrial towns in northern California were celebrating the hard-edged dancing and drinking music of the honky-tonks, where musicians had to compete with the amplified jukebox and the roar of the crowd. The new sound of country music had a raw energy and an imposing presence, which was enhanced by the piercing tones of the solidbody guitar. As Buck Owens said of it: "The sound just came about. I had a big old Fender Telecaster guitar, the walls of the buildings were hard, the dance floor was cement, the roof was sheet metal. There was considerable echo in there."[1]

The bands playing the dance halls of southern California were pushed into electrification more by the economic necessities of reaching the growing audience than by musical considerations—they needed more volume. Rose Maddox of the Maddox Brothers (and Rose) remembered, "After a while we got some amplifiers so we could be heard over the people."[2]

Country bands like the Maddox Brothers were rocking well before the 1950s. Their "New Muleskinner Blues" (1948) was based on a Jimmie Rodgers song, but it had all the energy and throbbing rhythm of rock 'n' roll: a turbo-charged, raucous, troublemaking stepchild of country. A year before Hank Williams had enjoyed a big hit with "Move It on Over," another up-tempo song with strong R&B influences and an electric guitar solo from Zeke Turner. Country acts thought nothing of picking up material in other genres. In 1956 the Maddox Brothers released a record called "The Death of Rock and Roll," which was a parody of an amateur singer's attempt at country vocalizing, based on Ray Charles's "I Got a Woman." (This in turn was taken from the gospel classic "There's a Man Going 'round Taking Names.") Rose said later: "I had got my songs and stuff in rhythm and blues, you know from the colored people's music. But I turned them around and did 'em my way so I thought I was doin' country music."[3]

This sort of confusion was common in the 1950s, before the business organizations running popular music had time to create race-specific categories to pigeonhole new music. Chuck Berry's first hit for the Chess label was "Maybellene," a country-influenced song (originally titled "Ida May") that crossed over into the pop and country charts.

In the late 1940s and early 1950s the electrified guitar sound had established its place in country music, although it shared the spotlight and the leads with the fiddle, the steel guitar, the accordion, and the mandolin. In the hands of West Coast players like Merle Travis and Jimmy Bryant, its sound was being articulated and refined. Country players were the intended market for the new solidbody electric, and these musicians' growing stature in pop music helped market this type of guitar. High-profile guitarists like Buck Owens and Jimmy Bryant provided valuable publicity for the manufacturers and demonstrated the flexibility of the new instrument. As a session man for Los Angeles's Capital Records, Bryant used his electric guitar to play a variety of country, jazz, and popular songs.

In the early days of the Telecaster (when it was called the Broadcaster), the Fender company was still a small business. Dealers from Los Angeles used to drive to Fullerton and pick up thirty or forty guitars and take care of the wholesaling, but in a short time the company had enough income from sales to build its own marketing network that would eventually cover the whole country.

In Memphis, Tennessee, Scotty Moore tried out an early Fender Esquire (the predecessor of the Broadcaster). He played in a local band called Doug Poindexter's Starlite Wranglers and did session work in Sam Phillips's recording studio, where he eventually met Elvis Presley and his musical destiny. Another acquaintance of Presley was also playing a new Fender solidbody at this time. Paul Burlison was playing gigs around town in Shelby Follin's Hillbilly Band. He later joined the Burnette brothers in the Rock 'n' Roll Trio and played his Telecaster on some records that ensured his place in rock 'n' roll history, such as "Train Kept a

Rollin'." Over in the west part of Memphis the young B. B. King tried a Broadcaster for his radio show on WDIA and his live performances. (This was years before he became associated with the Gibson ES-335.)

Down in New Orleans Eddie "Guitar Slim" Jones strapped on a Les Paul. In Shreveport, Louisiana, an up-and-coming session man called James Burton adopted a Telecaster. He used it for shows at the Municipal Auditorium that were broadcast on the *Louisiana Hayride* on station KWKH and heard over most of the eastern part of the country. When he landed the plum job of backing Ricky Nelson in the *Ozzie and Harriet* show, his Telecaster was seen on TV all over the United States.

In Chicago Muddy Waters tried out a Les Paul. He had come up from the Mississippi Delta and established himself in the vibrant music scene of the South Side. When he went into the clubs and bars, he started thinking about electrifying his guitar and getting an amplifier; in the noisy and competitive atmosphere of Chicago's music clubs, you needed something to be heard above the crowd. Everybody seemed to be playing louder than back home. Many blues players had added pickups to their acoustic guitars to get more volume, but this simple expedient did not give the musician the range of sounds, and the level of volume, that you could extract from a solidbody electric guitar. As Muddy Waters pointed out, the sounds from solidbody guitars were different from electrified acoustics—it was not just increasing the volume of the acoustic guitar; it was reaching out into a new world of sound. Waters was a major contributor to Chicago electrified blues, a genre that was to become increasingly influential as young guitar players rediscovered it in the 1960s "blues revival."

Not all the customers for solidbody guitars were destined to become major figures in rockabilly, blues, or rock 'n' roll; many of them played in jazz ensembles or backed popular vocal groups. In jazz circles the Les Paul had a stronger reputation than the Fender models—a measure of the status of Paul as a player and Gibson as a guitar manufacturer. There were still plenty of jazz players who adopted the Telecaster, most notably Oscar Moore of the Nat King Cole Trio.

It did not take long for one of the other big names in guitar manufacture to join the solidbody bandwagon. Fred Gretsch was disappointed in Gibson's move into solidbodies, because he felt it undermined the status and the craftsmanship of America's premier guitar makers who excelled in building archtops. Nevertheless Gretsch introduced models to compete with the Les Paul Goldtop. In 1954 the company introduced the Jet series, which bore more than a passing resemblance to the Les Paul. It also introduced a premium hollowbody electric guitar, the classic model 6120, and a solidbody model 6121. Both were endorsed by Chet

Atkins, a rising star in country and jazz guitar circles. At $230 the Duo Jet was priced close to the Les Paul, but the 6120 Chet Atkins was the top of the line at $385. The Gretsch electrics became the instrument of choice for rockabilly guitar players: Cliff Gallup of Gene Vincent's band used a Duo Jet and Eddie Cochran played a modified 6120.

At this time the Telecaster represented all of Fender's involvement in guitars. He had to contend with well-established guitar manufacturers who marketed a complete line of electrified instruments, including hard- and hollowbodied models. At the high end of the market there were two of the country's oldest and best-known manufacturers, and at the low end he was faced with low-budget models from Valco, Harmony, and Kay. It was only a matter of time before more cheap guitars entered the market. There was also the competition from individual craftsmen like Paul Bigsby, who were making custom solidbody guitars, including a teenager called Semie Moseley who made a double-necked guitar for Joe Maphis—the player who preceded James Burton in Ricky Nelson's backing band. The perfectionist in Fullerton, California, felt that he had to improve on his Telecaster and bring the solidbody concept to a higher plane.

The guitar he designed as the follow-up to the Telecaster was aimed at the same country audience: the cowboy-shirted pickers and steel guitar players who came out to the factory to visit or have their guitars fixed. It was their needs, as well as their music, that Fender kept in mind when he sat down at his bench and sketched out ideas for guitars and amplifiers. As he played with designs for the new model, he never could have imagined the impact it would have on popular music.

the stratocaster

The new model was much more than a refinement of the guitar that had done so well for Fender. It was entirely redesigned, with a different body shape and new electronics. In Leo Fender's mind it was a radical step forward in the technology of electric guitars. Comparing the looks of the Tele and the Strat is illuminating—one spare, functional, and workmanlike and the other refined, polished, and spectacular. One in plain blond wood and the other in dark-bordered sunburst finishes. The Tele was clearly made of wood—you could see the grain on the flat surface of the body—but the wood of the Strat was covered by layers of shiny paint that gave the solidbody a finish suggesting a plastic or even metallic surface. The Stratocaster had two radical cutaways, which gave the player access to the frets at the very base of the neck and gave the body a different look. Instead of the two pickups of the Telecaster, it came with three, which were activated by a three-way switch on the finger plate. The Telecaster had the bare minimum of controls in a functional straight line; its successor had a cluster of knobs and a switch flaring out from the strings toward the bottom of the guitar.

The tremolo system was touted as the major technological innovation of the new guitar. It certainly took the most time and money of all aspects of developing the model. Pulling or pressing the tremolo arm connected to the bridge brought a shimmering change in pitch. (It was in fact a vibrato unit, which changed the pitch of the strings, whereas a tremolo changes the volume. This misdefinition was enshrined by Fender in the years to come.) The tremolo unit was made as part of the "floating" bridge assembly, which had six separate pieces—one for each string. There were other tremolo units on the market: Beauchamp had designed one, and Doc Kauffman's and Paul Bigsby's versions had been adopted by several guitar manufacturers, but Fender thought he could do better. His unit produced excellent sustain and enabled the player to alter the pitch more than the competition's units—this was only a small consideration in the 1950s, but ten years later it was a major advantage when players used the tremolo arm excessively to get new sounds.

The Stratocaster was the work of several men. Like many of the innovations in the electric guitar, it was a collaborative effort of technicians and musicians. As the volume of work had risen in the factory, Fender had wisely sought out men to help him in the many different jobs he undertook. Freddie Tavares, a professional steel guitar player, was brought in as his personal assistant and collaborator on design work. Country musician Bill Carson was also hired on a part-time basis, and he made important contributions to the Stratocaster. Placing two musicians on the design team reveals how highly Fender valued their input and how much he depended on professional musicians to address technological problems.

The Stratocaster was designed to be easy to play for long periods. The contoured body with its beveled corners reduced the chafing on the player's body. Bill Carson is said to have suggested this feature, saying that the guitar should "fit like a shirt."[4] Bearing in mind that the great majority of Fender guitarists were country and western players—with tight-fitting cowboy shirts—this meant a snug fit between player and instrument. The Stratocaster does fit well into the player's body. Thinner, harder, and much more compact than the hollowbodied electrics, it allows the musician to be closer to the instrument and gives the feeling that the guitar is part of the player. It was also a whole lot easier to move around with the Stratocaster, making it ideal for the more animated playing in rock 'n' roll bands.

The jack-plug socket in most guitars was at the back or side of the body near the tailpiece, which often meant that the unseen cord got twisted or fouled. It invariably snagged on something if you attempted to sit down with the guitar in hand. On the Stratocaster the jack-plug socket is recessed into the front of the body (in a chrome-plated receptacle) at a forty-five-degree angle, making it easy to see and untangle the cord and keeping this vital connection out of trouble. This innovation was probably suggested by George Fullerton, the manager of the Fender factory, and it was patented in 1954 along with the tremolo unit and the six-section bridge. The Stratocaster presented a lot of new features for the money.

The shape and name of the Stratocaster, and its advertising copy, stressed newness and modernity; it was touted as "the most advanced instrument on the market."[5] The extra pickup on the Stratocaster was an example of marketing and technological advance coming together. When the electric guitar was first conceived, it was designed around only one pickup. The goal was to amplify sound, and the inventive effort went into the layout of the magnets and coils to achieve the most efficient conversion of motion into electricity. Guitar purists and some of the inventors claimed that one pickup was sufficient and adding more just muddied the sound. The first of the Fender line of solidbody guitars came with only one pickup. Adding more pickups was a concession to the musicians who liked the different tones coming from each pickup position: brighter and more metallic at the bridge and warmer and heavier toward the neck. Two pickups gave more versatility and a broader range of tones to choose from. The standard for electric guitars in 1954 was two pickups, so perhaps Fender put three on the Stratocaster to go one better.[6] He sometimes claimed that the choice of three pickups was determined by the large number of three-way switches in the factory, which were used for steel guitars and the Telecaster. Whatever the reason, the three-pickup combination was a desirable feature for musicians and a useful marketing tool.

The look of the guitar was both an extension of the Telecaster line and something completely different. The twin horns could have been inspired by tail fins of Cadillacs, or by the delta wings of jet airplanes. The Fender factory was not far away from several aircraft manufacturing plants, and this might have influenced the design of guitars. The prefix *Strato-* definitely had a resonance in the high-tech world of aviation. It was probably taken from the massive strategic bombers protecting the United States in the cold war at the time—the B-52 bomber was called the Stratofortress.

But in the midst of all this modernity there is a traditional western flavor to the design of the Stratocaster. The twin horns reflect a western motif, and it should not be forgotten that the sensuous curves on the guitar that Paul Bigsby built for Merle Travis look like the Stratocaster (but there is also some resemblance to the Les Paul shape). Western motifs were becoming very popular in country music in the 1950s. *Billboard* was now calling its country charts "Country and Western," and the Gretsch 6120 guitars were covered with western images—cacti, horseshoes, Indian spears, cow heads, and a "branded" G.

Whatever the inspiration, the shape of this new instrument broke the connection with the standard guitar. Unlike the many electrified acoustic models, it looked completely different from the traditional Spanish-style guitar. The Stratocaster shape came to define what a solidbody electric guitar should look like, and in the decades after its introduction in the 1950s, it has become the template for thousands of copies. Today there are numerous guitars reflecting the Stratocaster shape and layout.

the solidbody bass

Although the Stratocaster is the most influential solidbody guitar ever made, it would be difficult to argue that it changed the sound of popular music. It did not have that distinctive metallic twang of the Telecaster's bridge pickup, but it could do nearly everything the other solidbodies could do and then some, because its three pickups gave more tonal possibilities, especially when players found out that the three-way switch could be lodged between positions to produce different, hollow-sounding tones. A good player can make a solidbody guitar sound like a hollowbody, and vice versa. But of all the new electric guitars, there was one instrument that stood out immediately as a new electrified sound and a radical departure from the acoustic past.

The Precision Bass was meant to replace the acoustic bass, but it represented a completely new instrument with a sound that perfectly fitted the aspirations of rock 'n' roll (see plate 4). Of all the new electrified guitars produced in the 1950s it has the best claim to have changed the sound of popular music. A decade after its introduction, popular music such as soul or funk depended on the supple, powerful underpinning of the electric bass. Motown records in Detroit claimed to produce the "Sound of Young America" and succeeded in dominating American youth music in the years before the arrival of the Beatles. Influential Motown producers like Norman Whitfield made the electric bass an important part of their arrangements, and James Jamerson of the famed Funk Brothers rhythm section made it fundamental to the Motown sound. Band leaders like Sly Stone and George Clinton used it as a lead as much as a rhythm instrument. In the hands of players like Bootsie Collins or John Entwistle, it soon had a loud assertive voice of its own.

The electrified solidbody bass also brought some changes to the business of live music. The acoustic upright bass was a cumbersome, expensive instrument that suffered from the same low-volume problem as the guitar. Its bulk made it particularly unsuited to touring. Tied to the roof of the band's transportation (or filling up the rear seats), it rarely survived a tour unscathed. Here was a suitable problem for Leo Fender's inventive mind; his electrified instrument would be hardier and much more portable.

The first attempts to electrify the bass came in the 1930s. Lloyd Loar at Gibson designed one, and the Electro String company produced the first solidbody bass (designed by George Beauchamp) in 1935. A few other companies produced basses, and all of them were upright models paired with an amplifier. Fender approached the design of the bass from his previous work on the guitar, conceiving it as a solid-bodied instrument to be played like a guitar. It came with frets so that the player could enjoy more precise pitching of notes. The same electrical technology used in the Telecaster was applied to the bass. So was the shape; the bass was to be the big brother of the Tele.

It appeared on the market in 1951, a year after the introduction of the Broadcaster, and it had a strong resemblance to Fender's pioneer electric guitar. It had the same slab body and finish but was a little longer, with extended horns in a double-cutaway configuration; the left (upper) horn was longer and more curved in a shape that anticipated the Stratocaster design. It came with one single (four-pole) pickup with controls for volume and tone. The maple neck was bolted onto the body for ease of construction. The addition of an electric bass to a band brought a heavier dependence on electronic amplification. The Precision Bass was a part of a technological system that included amplifiers designed to handle heavier power loads and volume. (For the development of amplifiers, see chapter 6.)

The first important players to adopt the bass were from jazz. Oscar Moore, the guitarist with Nat King Cole, is reputed to have field tested a prototype, and Roy Johnson and Monk Montgomery (of Lionel Hampton's band) were early users who helped promote the product. Richard Smith points out in his book on Fender's guitars that the growing reputation of the Precision Bass gave the company a profile outside the West Coast and attracted jazz and pop players to Fender instruments.[7] Bill Black, who was in Elvis Presley's rhythm section, was playing an electric bass as early as 1955.

Fender basses and amplifiers created a new market for electrified instruments and so completely dominated it that *Fender bass* has become a generic term applied to all makes. Thus when one-time Motown songwriter George Clinton described the contributions made by James Jamerson to the Motown sound, he said, "James Jamerson was the epitome. He started Fender bassing. All that funk bassing, Jamerson was it."[8]

the electric guitar and rock 'n' roll mythology

If judged on the basis of how much it changed the sound of popular music, the Fender Precision Bass would have a much stronger case than the Stratocaster, yet it has none of the latter's prominence. How did the Stratocaster become an icon in American popular culture while the Precision Bass remains no more than a footnote?

The Stratocaster did not cause much excitement when it was introduced in 1954; there were plenty of new electrics on the market, and Gibson was still the company to watch. Many players and businessmen were taken aback at what they considered to be a bizarre shape for the guitar, with its long horns and massive white plastic pick guard, but it was not considered to be a revolutionary step forward in the technology. It took several decades for the Stratocaster to be hailed as the most influential solidbody guitar of all time.

The Stratocaster was the right product at exactly the right time. As rock 'n' roll reached epidemic proportions, country, blues, and R&B players were able to move out of the backwaters of popular entertainment, and they took their guitars with them. Although many of the great names played hollowbody Gibsons and Gretsches, the Stratocaster came to be so closely identified with the new music and its exciting young stars that it became a symbol of the movement. How could this have happened to an instrument that was not even around when Jackie Brenston recorded "Rocket 88"? It had a lot to do with the celebrity players who placed it within the broader context of the music and the mythology of rock 'n' roll stardom.

Like all the other guitar manufacturers, the Fender company used professional musicians to promote its products, so when Fender introduced his new guitar, he naturally approached the country and western guitarists who had adopted the Telecaster. But fortunately for him, a new type of country musician was picking up his guitars and using them to produce sounds that the inventor probably never thought of when he developed the Stratocaster in his workshop.

Buddy Holly was one of Fender's new customers. A teenage amateur musician, Holly was a product of the musical environment of West Texas, which was soaked in the sounds of western swing, bluegrass, and honky-tonks. He was also heavily influenced by the Texas blues and the R&B he heard on the radio on the *Louisiana Hayride* broadcasts. And he listened to Tejano music and country legends the Carter Family from the high-wattage radio stations broadcasting from the Texas border.[9] Holly's music was a amalgamation of all these styles, but he was also an innovator who produced pop songs that sounded fresh and appealing. He borrowed liberally from Chuck Berry and Bo Diddley (an example of the former: "Oh Boy!"; of the latter: "Not Fade Away"), yet somehow his voice and guitar sound had a directness and an innocence that made it stand apart. As a typical white teenager Holly played an important role in moving rock 'n' roll from what many considered to be dubious ethnic roots to the middle of the road. Frank Sinatra's famous diatribe about rock 'n' rollers being "cretinous goons" could hardly apply to the polite and cheerful Holly.[10]

Unlike most of the stars of pop music at that time, Buddy Holly wrote his own material and was determined to exert an influence in the recording studio, where he quickly mastered important new techniques such as double tracking. His only peers in this critical area were Chuck Berry and Eddie Cochran. Holly used the studio to thicken the sound of guitar, bass, and drums. Despite the layers of recorded sound, the ringing tones of his Stratocaster emerged as the trademark sound of his music: crisp and clear, its sound rang against the falsetto whoops and hiccups of his vocals. When Chuck Berry sang about his rock 'n' roll hero playing his guitar "just like a ringin' a bell," he might have been talking about Buddy Holly as well as thousands of other fresh-faced kids who were making music with Fender's new guitar.

Although for most of his career Holly played in front of drums and upright acoustic basses, his backing band eventually joined the trend toward electric bass guitars. Joe Maudlin made the transition to the Fender Precision Bass and the Bassman amplifier in 1958, citing its portability and convenience. Not only was it more comfortable to play; it also provided more volume to compete with the other electrified guitars in the band.

In Holly's hands the electric guitar assumed the roles of lead and rhythm instrument—a necessity forced upon him as sole guitarist in the group. When they went into a recording studio (and often on the road), the band brought in other guitarists, but in the early days Holly's guitar had to double as rhythm and lead. The electronic system under his control could produce enough amplified sound for a small band to come across as a big one, but Holly still had to cover the sounds and roles of two separate musical functions. Instead of the fluid single-note leads that guitarists like Les Paul had made their own, Holly relied on chords to cover both rhythm and leads. In the solo to "Peggy Sue" he used big A and D chords strummed hard, switching through the Stratocaster's three pickups to get a different tone when he moved from rhythm to lead. He could not move his hand to the switch quick enough without breaking rhythm, so he had someone work the switch for him as he played it in the studio. In using chords throughout the song Holly modified the existing rock song architecture of three verses of vocals followed by a single-note solo and then back to the chorus and the closing verse. The driving chord solos on "That'll Be the Day" and "Peggy Sue" rocked the song along while banishing the gentle virtuosity of guitarists like Les Paul to the sidelines.

The Fender Stratocaster that Buddy Holly purchased for around $250 at Adair's music store in Lubbock, Texas, in 1956 did a lot more than earn him a living. It made a statement about his involvement in professional music. He was using a Les Paul Goldtop at the time, but the Stratocaster had quickly become a favorite of country players in Lubbock, and Holly had to have one. His rationale for buying the guitar is significant: "I know good and well I can make it if only I had me a decent guitar and some decent clothes."[11] He wasn't just buying a guitar; he was embarking on a career as a professional musician. The Stratocaster was to accompany Holly to his first commercial recording session with a major record label. In the 1950s a chance to go to Nashville to record for the Decca company was a ticket to the big time, and Holly made two important business purchases when news arrived that Decca was offering him a recording contract: a set of loud stage clothes and a new guitar. Both were emblematic of his transition from high school part-time musician to recording star—just like his idol Elvis Presley.

Rejected by Decca's studios in Nashville, his career as a rock star was all washed up in January 1957. He tried again in Norman Petty's independent studio in Clovis, New Mexico, in February. His record "That'll Be the Day" quickly climbed up the *Billboard* charts while Holly was putting down tiles with his brother in Lubbock. By June 1957 his record had sold so many copies that the band was invited to New York City. In July their record label renewed their contract and they were booked on their first national tour. By August their record topped the charts and they appeared on the *American Bandstand* television show. If he had lived a little while longer, Holly would probably have made a movie in Hollywood. It would have been one of the rock 'n' roll movies that were popular

Buddy Holly playing a D chord on his Fender Stratocaster. From the collection of Rock and Roll Hall of Fame and Museum.

with teenagers at the time—a story about ordinary high school kids, gas station attendants, or truck drivers being transformed overnight into heroes of their generation. The shiny instrument slung around the shoulders of the rock star was a symbol of this transformation, along with other symbolic artifacts such as a pink Cadillac.

The mythological history of rock 'n' roll is basically a retelling of the traditional American rags-to-riches story. Throughout the nineteenth century the American dream of wealth, independence, and security leaned more toward the stroke of good luck than the gradual accumulation of riches—the sort of good luck that brought about the great inventions. But only a few could dream of being another Thomas Edison; for millions of nineteenth-century immigrants the fast track to success was professional sports or organized crime, not the world of technology. Hollywood colonized the dreams of many in the twentieth century with the magic wand of stardom, until rock 'n' roll took root in the minds of millions of teenagers.

Unlike the traditional stars of popular entertainment, Holly was a plain young man with thick glasses and crooked teeth, but it was this ordinariness that connected him so completely with his teenage audience. His rise to fame was as meteoric as that of Presley, but the fact that a gawky high school kid from West Texas could be a rock star was not lost on his fans, who realized that if he could, then they could. His biographer Philip Norman has pointed out that Holly's recorded guitar solos were an extended guitar lesson for thousands of amateur players all over the world who hoped to follow in his footsteps. Norman remembered, "The sound seemed to light up the whole of grey, boring Fifties Britain and my own grey, hopeless boyhood. Before that moment I had no ambition."[12]

In this context the electric guitar was more than a musical instrument. It was the tool to engage the business of popular entertainment. Chuck Berry's anthem to rock 'n' roll "Johnny B. Goode" describes an upbeat version of the American dream: a young boy leaves home with nothing but great ambitions and a guitar (carried in an old sack) and eventually reaches the top of the entertainment world (Hollywood) on the basis of his ability to play the instrument. This journey perfectly represented the American value of advancement through talent. The electric guitar was seen as the great enabler—taking poor kids from their dismal, ordinary backgrounds in remote parts of the country and propelling them into the big cities and the consciousness of the nation. It even worked for African Americans—one of the few consumer durables that did.

The guitar might have been the artifact of this process, but it was only a very small cog in an extended network of business and technology that made entertainment a global enterprise. The rise of rock 'n' roll stars is often viewed within the context of the music, but it is also a testament to the growing efficiency of business organizations and the powerful technological networks they commanded. Record companies, concert promoters, radio and television networks, and motion pictures all played their part. In the two short years that Buddy Holly spent in this business, his music and image achieved worldwide distribution.

Every performance he gave was a reenactment of the transcendental power of rock 'n' roll. It was also a showcase for Fender's new guitar, and even those kids who did not have the money for a ticket to the show (such as John Lennon and Paul McCartney in Liverpool) got the picture by word of mouth. In the austere postwar atmosphere of England, the promise of the new music shone through: almost anyone could do it—replicate the sounds, rise above the ordinary, and find fulfillment—and all for the price of an electric guitar and the effort of playing it.

Buddy Holly lived the life of a rock star and died a martyr's death on the road. His life story was commandeered as one of the great examples of rock 'n' roll stardom and sacrifice—a myth that enshrined the values of creativity and hard work (and some luck) within the rural-to-urban journey and the rise of the youth culture in America. The myth was important in the business of selling records. When the record company reissued Holly's catalog after his death, it was appropriately called *The Buddy Holly Story* (see plate 5). In the movie and the play about his life, the central branding image is of Holly playing the guitar; over time his image was fused with that of his guitar. He is so closely identified with the Stratocaster that it is hard to find a picture of him not holding one. Fittingly, Buddy Holly's gravestone in Lubbock is decorated with a bas relief of the guitar he helped make famous.

Sources

Buddy Holly Recorded Versions. Transcribed by Fred Socolow. Milwaukee: Hal Leonard, 1999.

Escott, Colin, and Martin Hawkins. *Good Rockin' Tonight: Sun Records and the Birth of Rock'n'Roll*. New York: St. Martins, 1992.

Scott, Jay. *Gretsch: The Guitars of the Fred Gretsch Company*. Fullerton, Calif.: Centerstream, 1997.

Notes

1. Nicholas Dawidoff, *In the Country of Country* (New York: Vintage, 1997), p. 242.
2. Ibid., p. 219.

3. Robert K. Oerman, liner notes to *Maddox Brothers and Rose* boxed set, Bear Family Records (Germany), 1998, p. 11.

4. Forrest White, *Fender: The Inside Story* (San Francisco: Miller Freeman, 1994), p. 80.

5. Richard Smith, *Fender: The Sound Heard around the World* (Fullerton, Calif.: Garfish, 1995), p. 141.

6. Gibson had produced a guitar with three pickups in 1949, the ES-5.

7. Smith, *Fender,* pp. 106–7.

8. James Waller, *The Motown Story* (New York: Scribner's, 1985), p. 158.

9. Interviews with friends of Holly in Lubbock, *Guitar Player,* June 1982, pp. 98–102.

10. Sinatra used terms like *brutal* and *ugly* to describe music he (and many others) thought was degenerate. See Gerald Early, *One Nation under a Groove: Motown and American Culture* (Hopewell, N.J.: Ecco Press, 1995), pp. 20–23.

11. Via Larry Holley, in Philip Norman, *Rave On: The Biography of Buddy Holly* (New York: Simon and Schuster, 1996), p. 73.

12. *Daily Mail* (London), 12 November 2002, p. 11.

Further Reading

Gordon, Robert. *Can't Be Satisfied: The Life and Times of Muddy Waters.* New York: Little, Brown, 2002.

Morrison, Craig. *Go Cat Go! Rockabilly Music and Its Makers.* Urbana: University of Illinois Press, 1996.

5 recording

the search for the sound

Susan Schmidt-Horning

Until 1948 high-quality recording involved cutting a live performance directly to wax- or lacquer-coated disks, recording from beginning to end with no simple means of correcting mistakes short of re-recording the entire piece. The playing time on a 78 rpm disk was only four minutes. However, despite a lengthy musicians' strike, 1948 proved auspicious for the recording industry and for the future of the electric guitar. Les Paul debuted his "New Sound" with a multilayered disk recording of "Lover"; Ampex Electric Corporation of California produced the first professional-quality tape recorders in the United States; Bell Telephone Laboratories announced the invention of the transistor; Columbia Records introduced the twelve-inch 33⅓ rpm Longplay record; and Leo Fender was designing what would be the first successful, mass-produced solidbody electric guitar. By the time Fender's guitars appeared in 1950, Decca, Capitol, Columbia, and RCA-Victor had converted their studios from disk to tape recording, editing and re-recording to correct performance flaws became common studio practice, and albums played for more than fifteen minutes on each side.

The developments of those first six months of 1948 laid the groundwork for a revolution in sound that occurred over the following two decades. Equipment manufacturers such as Magnacord and Presto soon introduced magnetic tape recorders to compete with the expensive Ampex Model 200. As affordable, professional-quality recording equipment became more widely available in the early 1950s, more enthusiasts entered the field of audio engineering, giving rise to small, independently owned and operated studios that challenged the recording hegemony of major record labels. The big record companies had the best recording equipment, experienced engineers to run and maintain it, and the record distribution networks that gave them control of the market. But they also had established methods of operating, union regulations, and professional standards to uphold. Independent studios, on the other hand, became low-tech laboratories of technological and musical innovation throughout the 1950s and into the 1960s. They may have lacked the technological muscle of a CBS R&D department, but in storefront studios across the country self-taught recording engineers and maverick producers explored the capabilities of the new technology of tape recording. They struggled to capture the loud and raunchy sound of rock 'n' roll, experimented with achieving new sounds, and provided professional recording facilities for undiscovered talent. Having cut their teeth in radio broadcasting or in the Army Signal Corps, or simply having dabbled in electronics since their youth, these budding audio entrepreneurs were caught up in a wave of postwar technological enthusiasm. Inspired by the music as much as the tools with which to capture it on record—and surely mindful of the potential for wealth and fame—their penchant for experimentation and entrepreneurship fueled the burgeoning recording industry of postwar America and the growth of rock 'n' roll.

The electric guitar became central to this new site of cultural production. By the 1950s new recording techniques made it possible for the guitar to move up front on record just as the amplifier had given it prominence on stage. The transition was often a recording engineer's nightmare and a creative producer's dream. When Guitar Slim recorded "The Things That I Used to Do" for Specialty Records in 1953, he waited until studio balances were carefully adjusted, then cranked his volume up to 10 once the tape rolled, redlining the VU meters. Chess Records owner Leonard Chess told Buddy Guy, "Ain't nobody gonna buy that noise, man," when Guy tried to persuade him to record the distorted-feedback guitar music his Chicago audiences loved.[1] In Memphis, Sam Phillips stuffed paper into a cracked speaker cone on Willie Kizart's amplifier, jury-rigging one of the first recorded fuzztone guitar sounds for Ike Turner and his Kings of Rhythm's 1951 recording of "Rocket 88." In Phoenix, Arizona, producer Lee Hazlewood spent days perfecting guitarist Duane Eddy's "million dollar twang" for "Rebel-'Rouser" by using an inexpensive microphone and speaker placed at opposite ends of a cast-iron storage tank in the corner of the parking lot at Ramsey Recording.

Through experiment and necessity, the recording studio became the launching pad for what has been aptly named "the search for the sound."[2] Not only was the electric guitar recorded and reproduced; it was reinvented—over and over again. For creating the sound of the electric guitar is no less a form of invention than designing its body or internal wiring. Les Paul may not have been able to patent his "New Sound" in 1948, but it became his sonic signature, achieved by manipulating recording technology to create an array of experimental sounds that were novel at this time. By the 1970s echo, reverb, fuzz, flanging, and phasing would become standard in the growing arsenal of effects employed by electric guitarists on stage and in the studio. But most of this had evolved—by accident as often as by design—from sounds produced within the recording studio. To understand that evolution, we need to look at how the guitar first became part of ensemble work in the studio.

becoming audible—recording the guitar, 1921–1939

Before guitarists could begin to think about shaping their sound, they had to be able to hear themselves. The guitar's immediate predecessor in jazz and big band music was the tenor banjo, an instrument far more piercing than the acoustic guitar. In 1921 banjo player Nick Lucas was the first to play guitar on record. Working with bandleader Sam Lanin, Lucas and the tuba player had to sit in the back of the studio at a safe distance from the acoustical recording horn, because the tuba's loud notes and the banjo's penetrating sound made the recording stylus jump off the wax disk, which ended the recording. These were the days of acoustic recording, when technicians had no way of adjusting the relative volume of any instrument after the record was made and could only achieve a balanced sound through careful placement of instruments around the recording horn. The horn collected the sound, which in turn moved a diaphragm that drove a cutting stylus to engrave the sound onto a heated wax disk. The louder the sound, the more powerfully it drove the stylus, sometimes forcing it off the disk. In the same way softer sounds like violins and guitars were not easily picked up by the horn. The first time Lucas brought his guitar to the studio, Lanin knew its sound would never be picked up from the back of the room, so he placed Lucas and his guitar directly under the recording horn. The results were a smoother rhythm and no trouble with needles jumping the grooves, and from 1921 Lucas played guitar rather than banjo on Lanin's recording dates.[3]

By the 1930s electrical recording and the use of microphones and amplifiers had long since replaced acoustic recording and the cumbersome, temperamental horn and diaphragm arrangement. The broader frequency response of recording systems and the greater sensitivity of microphones made it possible for the guitar's subtler tones to come across on record, as long as it did not have to compete with the other instruments. Early in the decade Eddie Durham had begun to play the resonator guitar with the Jimmie Lunceford Orchestra. As Durham later recalled, "Lunceford was crazy about the resonator. . . . He used to bring the microphone right up to the F hole of the guitar, so that between that and the resonator it was almost like having an electric instrument."[4] Durham's solo on Lunceford's "Hittin' the Bottle," recorded September 1935, has been credited as the first amplified guitar on record. Ironically, Durham was not primarily a guitarist but an arranger and trombonist who only doubled on guitar occasionally. Because of this he felt free to experiment with electrification when purists shunned it as a gimmick. In fact he felt compelled to amplify because, in his words, "The bands were *loud!* . . . I just knew that I had to be loud, too." Durham experimented with various alternatives to project his sound, including adding sound reflection devices to the body of the guitar and converting radios and phonographs into amplifiers. Some of his schemes posed a major problem for the primitive electrical systems in certain venues, and Durham later recalled that he "could blow out the whole sound system—even the lights—in the hall . . . in the middle of a gig!"[5] The distortion created by too much volume added an organlike quality to the sound, eliminating the distinct percussive effect of rhythm playing, so guitarists like Freddie Green, known for his clockwork-steady rhythm playing, never amplified at all.

Durham played the instrument more as an amplified acoustic than a solo instrument in its own right, and it was not until the late 1930s that Charlie Christian and T-Bone Walker began to exploit the full potential of the electric guitar. Christian had been inspired to take up the instrument by hearing Durham play in the spring of 1937, but his style was subsequently influenced by the playing of saxophonist Lester Young. Christian's playing transformed the guitar from a rhythm to a solo instrument and "gave the electric guitar such renown that almost all guitarists switched from acoustic to amplified instruments at the turn of the thirties."[6] In a December 1939 *Down Beat* manifesto, Christian put out the call for other guitarists to take up the electric and gain "a new lease on life."[7] By then even electronics enthusiasts and ham radio operators had taken an interest. A month earlier *Radio-Craft* featured detailed instructions from "amateur operator W9BMN" for building a "junkbox" electric guitar, and an earlier article explained how to build a magnetic pickup in order to connect the guitar to an amplifier and listen through a set of headphones, or "rattle the window panes, at that dance next month, by using a loudspeaker!"[8] Clearly, the concept of high volume did not originate with rock 'n' roll!

chasing the "new sound," 1937–1951

By the late 1940s, Les Paul had taken the electric guitar *and* recording techniques a giant step further. While most often associated with the instrument that bears his name, he contributed far more to refining the art of multiple (or sound-on-sound) recording than to the design of the Gibson Les Paul. He pioneered the art of inventing a signature guitar sound based primarily on recording technique. His "New Sound"—impossibly fast arpeggios, high-pitched, nervous-sounding tones, the percussive effects of muted rhythm chords, echo, and delay—evolved from painstaking experimentation, careful planning, and years of failed experiments. A childhood curiosity about the player piano, the phonograph, the radio, and the telephone in his parents' home became the basis for a lifetime of experimentation with sound. Paul vividly remembers, soon after learning to play the harmonica and the guitar, beginning to "crave" to hear what he sounded like, and in his words, "that's when I started to figure out how to record." He watched the way things worked, analyzed the mechanisms of these home entertainment and communication devices, and built his first recording lathe in the late 1920s. This was very primitive but workable enough to enable his mother (whom he calls "my first engineer") to record his performance of "Don't Send My Boy to Prison" when Paul played it on the Marquette University fifty-watt radio station, WHAD.[9]

During the 1930s Paul learned by observing radio engineers operate transcription machines and create novel sound effects in the Chicago and New York studios where he performed on soap operas and radio shows. Buying parts from electronics stores like Allied and Lafayette, he cobbled together disk recording equipment in his apartments in Chicago and New York. He began experimenting with primitive ways of laying down multiple parts on disk around this time, mainly because he wanted to record the arrangements he heard in his head and, he later explained, he "couldn't find a bass player or rhythm guitar player who wanted to play at three in the morning." Most of these recordings were, by his own admission, "terrible," and this early foray into multiple recording was a frustrating flop. He brushed it aside for the time being, but as he later recalled, "the light was lit."[10]

After moving to Hollywood in the 1940s, Paul installed a recording studio in the garage behind his house with the help of two friends and fellow electronics enthusiasts, Lloyd Rich and Vern Carson. He resumed his experiments in multiple recording in order to layer his melodic ideas on disk and to achieve a sound all his own. Paul was not the first to try sound-on-sound recording. In 1941 RCA recording director John Reid encouraged Sidney Bechet to overdub (or copy over) six instrumental parts on a disk after witnessing an oboe overdub on a symphonic record. But the process was time-consuming and each successive dub (or copy) added more degradation to the sound quality, muddying the parts that were recorded first. Paul discarded hundreds of disks before achieving the quality he was after, finally unveiling his "New Sound" in 1948: a multigenerational disk recording of "Lover" and "Brazil" featuring several guitar parts layered in careful order so that the parts he wanted most prominent on the final record were recorded last, enabling them to come through with greatest fidelity. The special effect of speeding up certain parts, by recording them at half-speed and playing an octave lower, or playing in the regular key, which was raised by the increased speed, made his guitars sound otherworldly and at times almost chimelike.

By 1950 Paul had begun to record multiple tracks using a modified Ampex 300 full-track (monophonic) tape recorder. This process proved trickier than the multiple-disk recording because it necessitated erasing the prior recorded track as overdubs were being added to it. With the success of the records he made with his wife and partner, vocalist Mary Ford, Paul was able to buy a prototype eight-track tape recorder from the Ampex Special Products Section in 1956, almost a decade before most recording studios adopted eight-track recording and before even stereo or three-channel recording was common. The Ampex eight-track recorder employed Sel-Sync (selective synchronization), which streamlined the process of multiple recording by allowing him to record the different parts in synchronization. But this method of recording, though it offered protection against losing prior tracks, lacked one key element that distinguished the earlier multiple recordings like "How High the Moon," namely the tension of having to get it right the first time, which made for a more spontaneous performance. In using the earlier mono recorder, the step-by-step process of playing along with a track while recording over it entailed starting from the beginning if a mistake was made on any subsequent track. There were no safety nets, and one could not "fix it in the mix" as became customary later. But while that involved greater risk, and thus required intense concentration and technical expertise, it also made for more creative tension, something the guitarist himself later admitted was lost with the multitracking method that enabled him to redo parts ad infinitum.

Les Paul's "New Sound" signaled a conscious attempt to create a unique musical identity in the studio, which the record company eagerly grasped as a promotional vehicle. As the album jacket for one of his Capitol releases proclaimed,

Les Paul in his home recording studio, surrounded by the things nearest and dearest to him: his namesake guitar (heavily modified), his tape recording machine, and his wife, Mary Ford. Courtesy of Thomas W. Doyle and Tom Wheeler.

"The search for a hit record is the search for 'a sound,' that unique combination of an artist's individuality, imaginative musical arrangement, and *skillful engineering.* . . . no one has been as successful in creating hits as Les Paul."[11]

Les Paul was a pathbreaker in the electronic manipulation of sound within the genre of popular music, an achievement derided by serious jazz contemporaries yet recognized as pioneering by musicians and recording engineers decades later. At a time when singing any closer than several feet from a microphone was considered against the rules of proper studio recording, he devised close-miking techniques for vocals as well as for guitar, in part to reduce unwanted street noise that bled through the walls of his Hollywood garage studio. He also devised tape-slap echo techniques that were later copied by others. But he was by no means the only one tinkering with equipment, experimenting with technique, and inventing electronic gadgetry, or as he put it, "chasing sound."[12] While Paul has been credited with a number of technical innovations, these achievements were often the result of collaborations with manufacturers and engineers who implemented his ideas.

Paul chose to record at home for several reasons. First, it gave him the freedom to experiment endlessly, a luxury that few other musicians enjoyed or desired. Second, recording at home enabled him to retain control of his creative process and outcome: he did not have to defer to record company staff engineers' requirements and producers' wishes. Finally, it afforded secrecy of methods, which for an artist like Paul were nothing short of proprietary; for even he would admit that his distinctive sound was less a result of a unique playing style than the way he used recording technology to fashion his sound. If others knew how he achieved it, they could do the same. By protecting his methods, he ensured that his home studio recordings of the 1940s and 1950s would remain both technical and musical marvels, unique and absolutely visionary. Here was an established artist recording for a major record label who never used the company studio and avoided even coming to the Capitol offices to play his masters, choosing instead to phone them in for approval. In the larger context, Les Paul's determination to push the capabilities of the equipment at hand embodied the spirit of technological enthusiasm that infused studio recording in the postwar period.

capturing the expression, the total effect, 1951

While Les Paul was in the midst of developing what he called his "electronic gadgets" and the technique of multiple-disk recording, the genre of music that would soon take over the recording industry was emerging from small independent studios that employed much cruder recording technology. Amplification had enabled the guitar to hold its own with the rhythm section in clubs and on record. In fact in the days before the electric bass became widely adopted, guitars often overpowered the acoustic upright bass, inspiring both ingenious miking methods and at least one short-lived invention, the amplified bass ocarina, invented specifically to meet the need for a louder bass sound for broadcasting and recording. But amplification also introduced yet another technological variable, bringing potential headaches and unforeseen problems like blown tubes or cracked speaker cones. Nevertheless, such technological glitches could be translated into innovative sounds with the right engineer at the controls.

Sam Phillips was one such recording engineer. Originally a radio announcer who began recording as a sideline, Phillips not only had an ear for talent (he discovered Elvis Presley, among other performers); he was equally savvy and open-minded about what made for a good sound in the studio. He made the most of his recording equipment, which in the early 1950s (before he acquired professional Ampex tape recorders), included a Presto portable tape recorder, a mixer, a cutting lathe, and a turntable. In terms of his approach to recording, Phillips's motivation came from "the freedom we tried to give the people, black and white, to express their very complex personalities." Phillips had learned early on to adapt to any contingency in the recording environment and to keep his ear attuned to what he called "the *expression* . . . the total effect." Thus when Ike Turner's band arrived at his studio after a drive from Clarksdale, Mississippi, during which the guitar player's amp had fallen from the roof of the car, Phillips was undaunted by the fuzz and static emanating from the cracked speaker cone. With no way of getting it fixed before the scheduled recording session, he started fiddling with it, stuffing paper into the rupture, and voila! In Phillips's words: "It sounded good. It sounded like a saxophone. And we decided to go ahead and record."[13] The result was not only Phillips's first hit recording but a number one song that became the second-biggest R&B record of 1951, a song written by Turner's sax-playing cousin, Jackie Brenston, about his car, the Oldsmobile "Rocket 88."

Recorded in March of 1951, the same month Capitol Records released Les Paul and Mary Ford's "How High the Moon," Brenston's "Rocket 88" portended not only the stylistic direction of popular music but perhaps the quintessential electric guitar effect of the 1960s, an effect the Yardbirds, the Rolling Stones, and other rock groups would exploit fully by the middle of that decade—the distorted fuzztone. But whether or not "Rocket 88" was the first recorded fuzz sound is not really significant; what is important is how the recording engineer—in this case Sam Phillips—adapted to the unexpected in the studio. The late music historian Robert Palmer noted that just about any other engineer at the time would have thrown a band with such broken-down equipment out of his studio. But Phillips transformed misfortune into opportunity. He was a different breed of recording engineer from the "control men" at major labels who sought pure, undistorted sound. For Phillips "the expression was the thing."[14] Because of that, he figures prominently among the first of the independent producers who were beginning to usurp the role previously held by industry artist and repertoire men (few were women then) who were on staff to sign singers and vocalists, choose the material they would perform, hire musicians and arrangers, and produce the records in the company studios. In fact Phillips appears to have liked the jury-rigged fuzz sound enough to intentionally repeat it. Within a few years he devised a method of reproducing his serendipitous discovery in a way that did not necessitate the permanent destruction of a speaker cone—by placing cardboard boxes over the amps and cutting a hole in one of the boxes to achieve a crude fuzz effect from the rattling sound. The late Carl Perkins witnessed this during his second recording session with Phillips in 1954. As Perkins recalled, Phillips made the most out of "low-tech gear; sheer ingenuity; and necessity, the mother of invention. . . . Sam was a good sound man. He'd fooled with it for a good while before he started making records with us, fooled with it with blues people, and he knew about recording."[15] Phillips embodied the pioneering, inventive, spit-and-sealing-wax approach to making records that characterized these independent studio operators who were not forced to adhere to company rules about permissible levels of volume or distortion. He was a consummate professional, having honed his engineering expertise the way recording technicians had done since the early days of the phonograph: by trial and error.

Feedback, distortion, and overdriven amps were integral to the styles of blues players like Guitar Slim and Muddy Waters, who had been cranking up the volume in clubs since the early days of amplification. Slim, in fact, opted to run his guitar through a cheap PA system instead of an amplifier specifically *because* it distorted so much. When he recorded a song written by Ray Charles, called "The Things That I Used to Do," at Cosimo Matassa's Studio in New Orleans, the producer, Johnny Vincent, reportedly insisted he play at the loudest possible volume to cover his somewhat sloppy timing and intonation, long masked by the low fidelity of his amplification system and the noisy clubs in which he played.[16] As the fidelity of recording technology improved, the recording studio revealed inadequacies in a player's technique, even as it offered the opportunity to cover up such weaknesses in performance through editing or electronically altering them.

By early 1958 Les Paul and Mary Ford had made their last recordings for Capitol Records. Now feeling outmoded by the rise of rock 'n' roll, Paul was nevertheless an influence, in substance if not style, on the rash of rock instrumentals that had begun to hit the market two years before with Bill Doggett's "Honky Tonk." The guitar had been a feature instrument on country and western swing records as well as jazz, dating back to the days of the popularity of the Hawaiian guitar in the 1920s, but 1958 gave birth to a new phenomenon, the guitar instrumental. By the early 1960s the golden age of the rock instrumental was in full swing, with the Chantay's "Pipeline," the Surfaris' "Wipeout," and a string of hits by the Ventures.

the search for the sound, 1958

One of its pioneers was Frederick Lincoln Wray, a country picker who backed up cowboy stars like Lash LaRue, Wild Bill Elliot, and Sunset Carson at outdoor fairs in the 1940s. In 1958 his group Link Wray and the Raymen recorded a song that featured by far one of the most menacing guitar sounds ever heard. Having achieved it initially by the accidental distortion that resulted from overdriving his amplifier and miking it closely on stage, Wray found that he had to punch holes in his amp's tweeters to get the same effect in the studio when the group recorded "Rumble." Within the genre of rock 'n' roll recordings, Link Wray's "Rumble" was considered "a daring move in 1958" and was banned from certain radio playlists for being "too suggestive." At a time when most professional guitarists did whatever they could to avoid distortion, Wray chose to capitalize on its brash and undisciplined power and in so doing managed to capture "the essence of adolescence."[17]

On the heels of Link Wray came a young western swing guitarist named Duane Eddy, who, with the help of former disk jockey Lee Hazlewood, made a career out of his twangy sound. In a small studio located behind a barber shop on North Seventh Street in Phoenix, Arizona—known as Ramsey Recording—this producer-artist team conducted the "search for the sound" in earnest, recording a succession of guitar instrumentals between 1958 and 1960 that inspired a generation of players. Along with a group of dedicated local session musicians (including one woman, Corki Casey), Eddy and Hazlewood consciously sought to *shape* a distinctive sound within the confines of the recording studio. Like Sam Phillips, Lee Hazlewood went for a total effect, but it was an effect generated by the electric guitar as centerpiece, as the source of what Hazlewood would christen, in a clever marketing move, "the million dollar twang."

Like others before him, Hazlewood managed to achieve the sound he wanted with some fairly primitive recording technology. The Phoenix studio had grown out of owner Floyd Ramsey's radio repair business and was primarily used to record commercials. The four-pot recording console was so small that they used a large cigar box as a dust cover, and without the benefit of an echo chamber Hazlewood's initial efforts in the studio involved hours upon hours of experimentation. He later remembered that when he recorded his first hit record, "The Fool" by Sanford Clark, "Connie Conway played 'drums' with one brush that we found in the studio, and a Campbell's tomato soup box over a drum stool . . . but [it took] nearly three weeks, on and off, to get Sanford's voice on it. It wasn't his fault, it was just me trying to get a sound on it 'cause we didn't have echo chambers so we tried all kinds of combinations of 7½ and 15 ips tape echo on little machines, stretching them and plugging them into each other 'til we got something like I wanted it."[18]

The first Eddy-Hazlewood collaboration was a song called "Movin 'N' Groovin," which became a chart hit for Jamie Records, featuring "Duane Eddy and His Twangy Guitar" on the label credit. Hazlewood tried to persuade Ramsey to invest in the addition of an echo chamber, but the studio owner refused, saying it would cost too much. So Hazlewood spent a day shopping around, yelling into storage tanks until he "finally got one where the yell came back to me [and] I said, 'right, we'll buy that one.'" He paid two hundred dollars for that cast-iron grain storage tank, moved it to the corner of Ramsey's parking lot, and placed "a four dollar mike at one end and a sixty cent speaker at the other and that's not exaggerating too much about the value of it."[19] Of course, this meant they could not record if it rained and they frequently had to shoo the birds or curious neighborhood kids away, but the cylindrical grain tank became an enabling technology for the guitar sound on "Rebel-'Rouser," their second record and the one that defined the classic Duane Eddy sound.

In fact few purposely built echo chambers existed in those days outside of radio broadcasting and film industry studios. Essentially an artificial, and thus controllable, means of generating the effect of performing in a large concert hall, the key features of echo (or reverberation) chambers are that they should be long and narrow and have extremely hard and smooth walls to reflect sound. Hence some of the best-sounding echo on records was achieved by using concrete stairwells, tiled washrooms, and even sewer pipes. Hazlewood eventually overdubbed a saxophone part, hand claps, and vocals at Gold Star Studios in Hollywood, which was famous for its deep echo chamber and was one of the studios in which producer Phil Spector would later create his signature "Wall of Sound" on

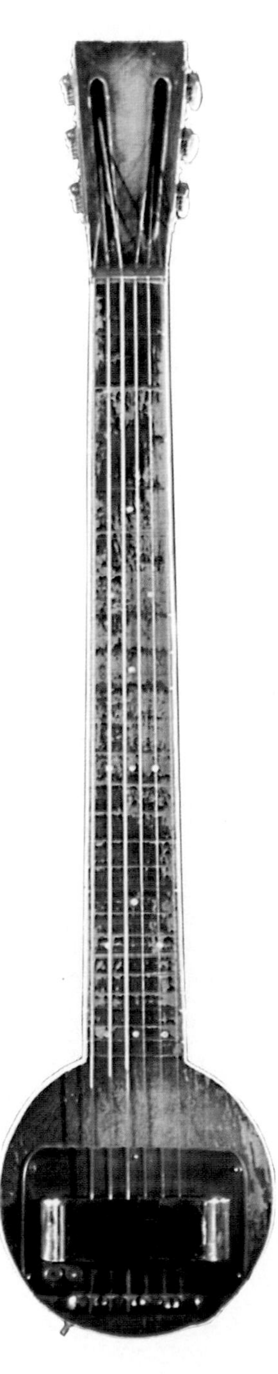

Plate 1. Guitars as art. A stunning blue guitar from the Chinery Collection that was featured in the "Blue Guitars" exhibit at the Smithsonian. This one was crafted by Steve Grimes in Hawaii. It is called Jazz Laureate and looks so good you might think twice before playing it. Courtesy of The Chinery Collection.

Plate 2. The Frying Pan in its production form. The Electro String Company's Electro Hawaiian guitar, circa 1931. Can this really be the first electric guitar? Photo by Richard Strauss, Smithsonian Institution, courtesy of Rickenbacker International Corporation.

opposite
Plate 3. Les Paul with his Log. Two halves of an acoustic guitar have been attached to each side of the Log. Photo by Richard Strauss, Smithsonian Institution, no. 96-4741.

Plate 4. The Fender Precision Bass, 1955 vintage. Smooth and supple—the power behind the groove. Photo by Richard Strauss, Smithsonian Institution, courtesy of The Chinery Collection.

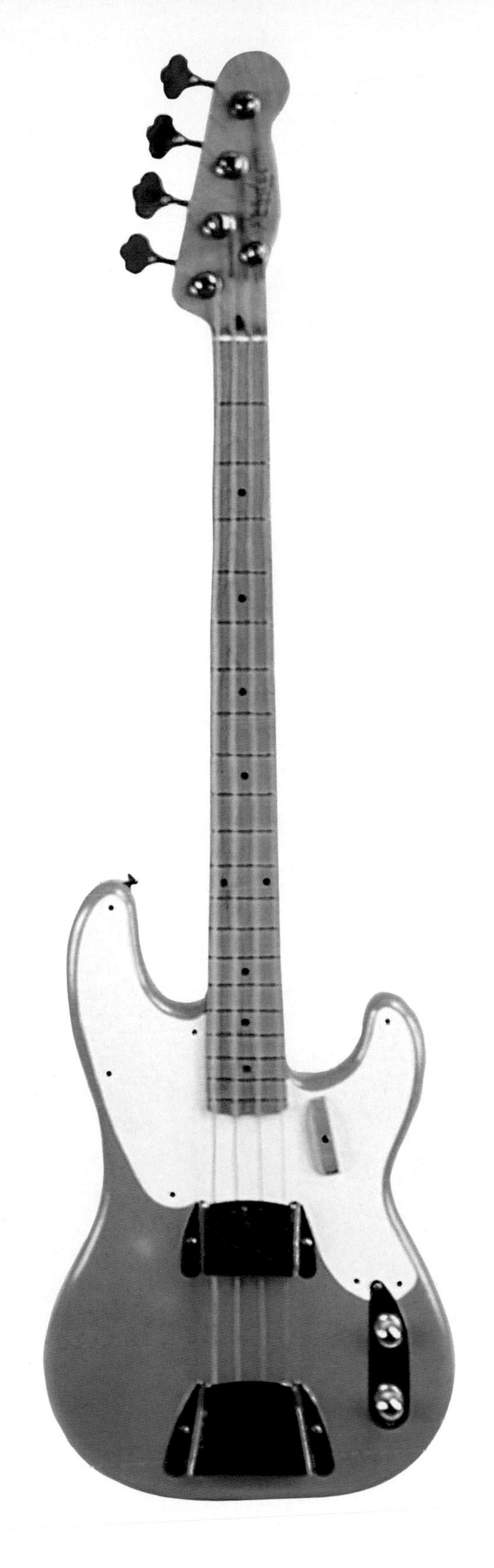

Plate 5. The long-running musical *The Buddy Holly Story* in London's theater district. Like the music, the image of Holly and his guitar lives on and on. Photo by André Millard.

Plate 6. *Left,* The Gibson Flying V. Not so much a guitar as something you might find on an alien spaceship. Photo by Richard Strauss, Smithsonian Institution, courtesy of The Chinery Collection.

Plate 7. *Above,* One of the first Fender Broadcasters off the line (number 27) with its tweed amplifier. Beautiful shape and beautiful color—a pin-up photograph for every guitar fan to admire. Photo by John Peden.

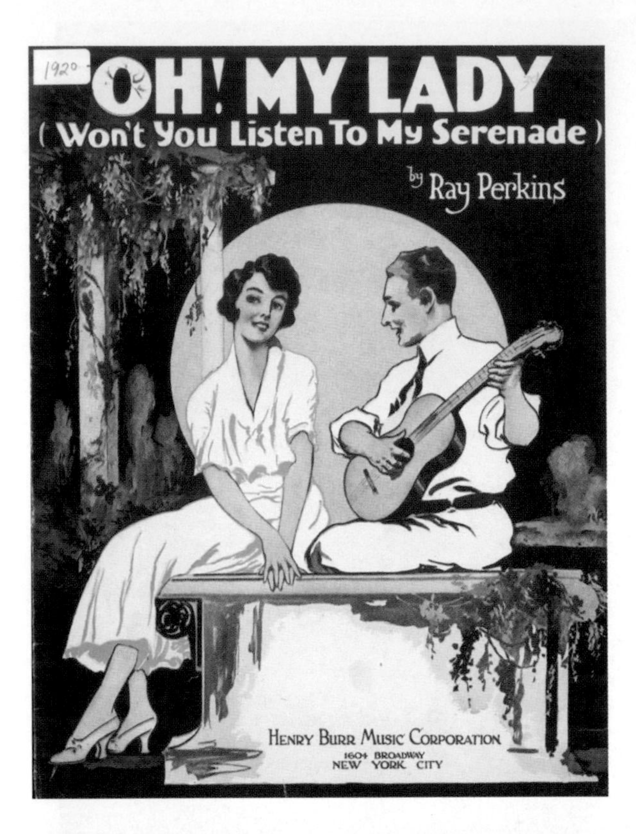

Plate 8. "Oh My Lady" sheet music. The old romantic view of the guitar—sedate, sophisticated, and eminently presentable. Courtesy of University of Colorado at Boulder, Howard B. Waltz Music Library.

Plate 9. When the fat lady finally sings, the electric guitar still drowns her with noise. The new look of the guitar—loud, obnoxious, and garish. Cover by R. Sikoryak, © 1996. Courtesy of R. Sikoryak/*The New Yorker*, August 26/September 1996. Reprinted with permission.

Plate 10. Eddie Van Halen playing his Frankenstrat guitar. The sticky tape underlines the fact that this guitar was constructed by the player. Photo by Neil Zlozower.

Plate 11. Prince with an extravagant custom guitar to complement his extravagant dress, stage show, and musical ideas. Courtesy of Paisley Park Enterprises.

Plates 12 & 13. The changes that thirty years and heavy metal wrought: Les Paul (*top*), Angus Young from AC/DC (*bottom*). Can you spot the differences? Les Paul image courtesy of Michael Ochs Archives. AC/DC is a registered trademark of Leidsepleine Presse B.V. Photo by Michael Putland.

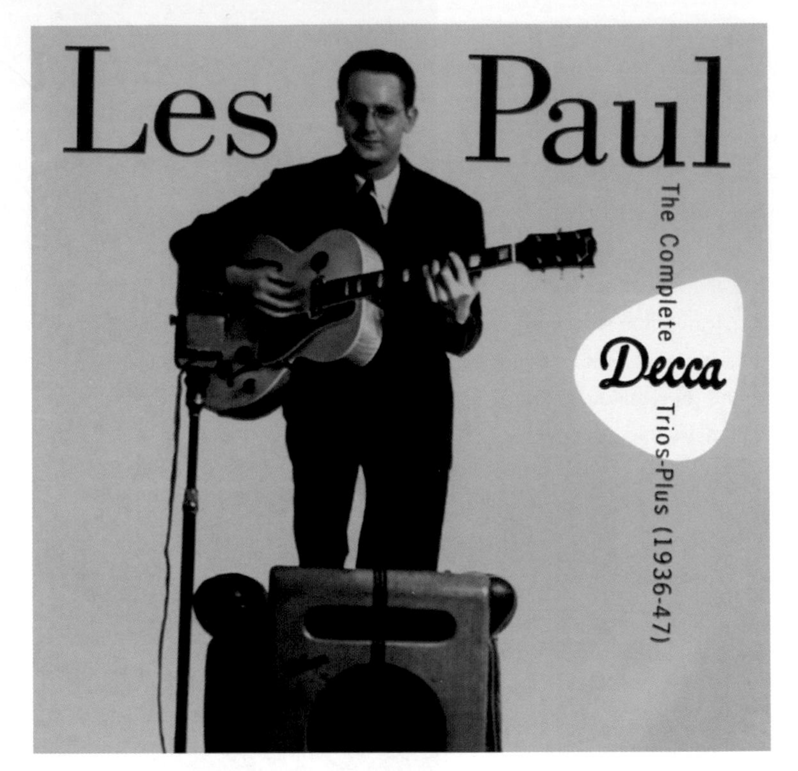

records by the Righteous Brothers, the Ronettes, and Ike and Tina Turner. Spector, who has been called the architect of sound, in fact learned many of his tricks observing Lee Hazlewood at Ramsey's studio. Hazlewood's partner, Lester Sill, brought Spector to Phoenix for a time. Sill recalled Spector's informal apprenticeship: "Phil would just sit in the studio. Lee really was a master at sound at that time, particularly with tape reverb and for hours he would sit and experiment in that studio with sound, and Phil would just sit and watch and listen for hours and hours—all the time he was absorbing, wouldn't utter a word. On occasion he would lean over and ask . . . 'Why are you now using 7½ echo or tape reverb?' Or 'Why are you just using plain echo rather than tape echo?' and so forth." The way Hazlewood remembers it, "The combination of echoes we used to play with so much with Duane used to drive Phil crazy—he wanted to know what everything was. I used to beg Lester Sill to get him out of the studio."[20] Hazlewood's reluctance to act as a tutor was more the result of a desire for concentration than a need to maintain secrecy, just as Spector's desire to learn the tricks of the trade was rooted more in curiosity than greed.

Hazlewood and Eddy continued to record a number of twangy instrumentals, all at Ramsey's studio. By 1959 they were able to help Floyd Ramsey upgrade his equipment, making it possible for the studio owner to raise his rates from eight dollars per hour (a low rate, which certainly enhanced the studio's appeal to Hazlewood) up to twelve dollars per hour. With Dick Clark as his manager, Duane Eddy's career skyrocketed. By 1962 Eddy signed a recording contract with RCA Records and was assigned a staff engineer, Al Schmitt, who was never quite able to achieve the same results in the sophisticated RCA Los Angeles studios that Hazlewood had obtained with the grain tank echo chamber in Phoenix. Schmitt admits to having difficulty communicating with Eddy, and he felt that Hazlewood had been the main force in the collaboration, able to get something out of the guitarist that no one else could. Eddy had grown accustomed to the loose atmosphere and cheap rates of the Phoenix studio that allowed for virtually endless experimentation and found it difficult to adapt to the major record label approach, which in 1959 made no allowances for artistic idiosyncrasies. At that time the standard studio session lasted three hours and yielded four songs; anything beyond that required overtime pay for union session musicians, who had to be proficient sight readers to maintain steady work in the highly competitive business of studio recording. Most performers wanted to go to the studio, make a record, and get out as quickly as possible. As a former engineer from the RCA Los Angeles studios recalls, "nobody would dream of spending more than two or three days recording an album. . . . you'd be laughed out of the business."[21]

The fact was that studio rules were changing—one might even say disappearing—along with the changing styles of popular music and the introduction of new recording technology. Multitrack recording had become available, and the Beach Boys and the Beatles had demonstrated what could be done by using the studio to craft concept albums. By 1967 Al Schmitt found himself recording the Jefferson Airplane, who drove motorcycles into the studio, brought in tanks of nitrous oxide and various other controlled substances, welcomed a steady stream of visitors into the studio, and often appeared hours late for sessions. The recording industry had changed dramatically from the time when Sam Phillips opened his Memphis Recording Service. Major record companies had absorbed the smaller independent labels and signed up the once unknown musicians on whom people like Phillips and Hazlewood had gambled a decade earlier. Eventually, even the major record companies began to indulge their increasingly experimental artists with virtually limitless studio time, often agreeing to let them record in the studios of their choice. Independent studios attracted musicians who sought a looser atmosphere that allowed them the time to experiment and provided an engineer willing to accommodate their ideas and perhaps even let them operate some of the controls.

hendrix: creating the sounds in his head

Jimi Hendrix did more to reinvent the sound of the electric guitar than any musician since Les Paul. Not surprisingly, Hendrix considered the recording studio his "laboratory," and he left behind hundreds of reels of tape from sessions in more than a dozen studios comprising "unfinished ideas, spontaneous moments of inspiration, and countless hours of unrealized ideas." Like Paul, Hendrix was impatient with having to wait for musicians to become available to execute his ideas for songs. As he later recalled: "If I write something about three or four in the morning, I can't wait to hear it played. It's even a drag to have to wait for the other cats to arrive."[22] Hendrix was fortunate to flourish at a time when independent studios were cropping up everywhere. Late-night sessions were not unusual, and after the mid 1960s rock bands rarely adhered to the three-hour, four-song expectation that guided earlier professional recording sessions.

A cultural as well as technological shift in studio recording had occurred, which gave Hendrix a tremendous amount of creative freedom and a broader sonic palette to work from in the studio. Whereas Les Paul layered melody upon melody, using changing speed, tape echo, multiple delays, and microphone placement, Hendrix experimented with controlled feedback from amplifiers, fuzztone, wah-wah pedals, and whammy bars (see chapter 6). He incorporated these with studio techniques such as echo, backward overdubbing, panning, use of equalizers (filters), and phase shifting to shape the sound of his guitar to conform with sounds he heard in his head. He liked to describe sounds as colors, as natural sounds or physical sensations, and the phenomenal success of his recordings of the late 1960s made it possible for him to experiment endlessly in recording studios from London to New York to Los Angeles to achieve these sounds. He spent so much money in studio time that his business manager finally persuaded him to invest in a studio of his own. Although not completed until the summer before Hendrix died, Electric Lady Studios provided him with the freedom to record, rehearse, experiment, and just hang out whenever he chose.

Essential to Hendrix's ability to continually reinvent his sound was the collaborative nature of studio recording, which had evolved in large part because of the precedent set by independent operators like Sam Phillips and Lee Hazlewood and engineers like Tom Dowd of Atlantic Records. Once Dowd, for example, explained the possibilities offered by multitrack recording to songwriters Jerry Leiber and Mike Stoller, they began to write their songs with the overdubbed parts in mind. For them the technology inspired a new approach to songwriting. Conversely, Hendrix described to his engineers what he heard in his head, or even what he simply *imagined* his guitar should sound like, and inspired them to come up with the means of achieving those sounds. Hendrix's biographers observed the phenomenon that began to sweep through the recording industry: the ready and willing attitude of recording technicians to accept any challenge to get a sound. "It was pure delight for creative studio technicians to hear somebody say, 'I want the sounds of underwater,' or be asked to recreate intergalactic warfare." Thus, in Hendrix's case, the musical idea provided inspiration for the electronic gadget. George Chkiantz devised a method of creating a phasing sound that Hendrix had heard in a dream by copying an effect used by the Beatles on their *Sgt. Pepper* album, and Roger Mayer spent a lot of time building and tweaking various fuzz boxes to Hendrix's specifications. And while the recording studio may have been, as Shapiro and Glebbeek noted, "the centre of his creative universe," Hendrix was as adamant about recreating the power of his live performance on record as he was about recreating the recorded effects on stage.[23]

In the early years of sound recording, records were evaluated on how faithful they were to the original performance. Record companies sought every means at their disposal to improve the fidelity of records, to recreate the original performance as closely as possible. With the introduction of electrical recording in 1925, that goal could at last be achieved, but only by manipulating frequencies and effectively altering the original sound. Because of the many complex variables involved in transducing acoustic energy into electronic signals, then converting it back to acoustic energy, staying faithful to the original performance was not all it appeared to be. With records like Les Paul's and Duane Eddy's, recording had made the transition from being a faithful reproduction of a live performance—the goal in Thomas Edison's day—to a studio creation that would only later be recreated on stage.

By the 1970s the electric guitar had assumed its place as *the* solo instrument in rock. Multitracking, punching in, editing, digital delay, equalization, phasing, and flanging, as well as all manner of signal processing devices had become standard tools, enabling guitarists to become consummate perfectionists in the recording studio. With studio effects even mediocre players could sound polished, leading many to take up the guitar because of the perceived ease of learning how to play it. Various forms of guitar synthesizers appeared in the late 1970s, and by the 1990s multi-effects systems, once only available as custom-built units, appeared on the market. In 1995 Roland introduced its VG-8 system, a complex signal processor that enabled a guitar player, via a set of digital "models," to mimic hundreds of possible sounds made by a guitar, effects, amplifiers, and speakers, and any combination thereof, without the use of amplifiers. Over a seventy-five-year period, the guitar had evolved from a barely audible rhythm instrument to a vessel through which almost any sound imaginable could be reproduced. The technological tools that made this transformation possible emerged from recording studios, either through accidental discovery or painstaking cut-and-try methods. Before the shortcuts existed, however, maverick musicians like Les Paul and engineers and producers like Sam Phillips, Lee Hazlewood, and countless others relied on a spirit of invention, adaptability, and dedicated enthusiasm in their search for the sound of the electric guitar.

Sources

Duane Eddy: Twangin' from Phoenix to L.A. Bear Family Records. 1994. Boxed set.

McDermott, John, with Billy Cox and Eddy Kramer. *Jimi Hendrix Sessions.* Boston: Little, Brown, 1995.

Sallis, James, ed. *The Guitar in Jazz: An Anthology.* Lincoln: University of Nebraska Press, 1996.

Notes

1. Robert Palmer, "The Church of the Sonic Guitar," *South Atlantic Quarterly* 90 (fall 1991): p. 666; Chess quote cited in *The Electric Guitar: An Illustrated History,* ed. Paul Trynka (San Francisco: Chronicle Books, 1993), p. 42. In a 1993 interview, Guy also recalled Chess refusing to let him bring a wah-wah pedal into the studio. Jas Obrecht, "Buddy Guy," in *Blues Guitar: The Men Who Made the Music,* by Jas Obrecht, 2d ed. (San Francisco: Miller Freeman, 1993), p. 210.

2. James Gang drummer Jimmy Fox, phone interview by author, 29 October 1996.

3. Nick Lucas and Jas Obrecht, "Nick Lucas," in Sallis, *The Guitar in Jazz,* pp. 12–19.

4. Eddie Durham quoted in Leonard Feather, "The Guitar in Jazz," in Sallis, *The Guitar in Jazz,* pp. 4–5.

5. Eddie Durham quoted in Stan Britt, *The Jazz Guitarists* (Poole, Dorset, U.K.: Blandford Press, 1984), p. 12.

6. Joachim E. Berendt, "The Guitar," in Sallis, *The Guitar in Jazz,* p. 149.

7. Christian quoted in Walter Carter, *Gibson Guitars: 100 Years of an American Icon* (Los Angeles: General Publishing Group, 1994), p. 163; and in Trynka, *The Electric Guitar,* p. 12.

8. Frank E. Chambers, "How to Build a 'Junkbox' Electric Guitar." *Radio-Craft,* November 1939, pp. 271, 308–9; Kendall Ford, "A Home-Made String-Music Pickup," *Radio-Craft,* April 1939, pp. 601, 624–25. Thanks to Alex Magoun for alerting me to these articles.

9. Les Paul, interview by author, New York, N.Y., 19 January 1999. Other accounts of this performance give the radio station as WTMJ, the *Milwaukee Journal* station.

10. "The Wizard of Waukesha," *Guitar Player,* February 1970, p. 17; Les Paul, interview by author, New York, N.Y., 9 February 1999.

11. Album liner notes for Les Paul, *The New Sound,* Capitol, 1950, emphasis added.

12. Les Paul, "Portraits of Invention: An Evening with Les Paul," 13 November 1996, Baird Auditorium, National Museum of American History, Washington, D.C.

13. Sam Phillips quoted in Robert Palmer, "The Church of the Sonic Guitar," *South Atlantic Quarterly* 90 (fall 1991): p. 658.

14. Palmer, "Church of the Sonic Guitar," pp. 658–59.

15. Carl Perkins, with David McGee, *Go, Cat, Go: The Life and Times of Carl Perkins, The King of Rockabilly* (New York: Hyperion, 1996), pp. 121–22.

16. Donald J. Mabry, "The Rise and Fall of Ace Records: A Case Study in the Independent Record Business," *Business History Review* 64 (autumn 1990): p. 427.

17. Chris Gill, "Profile: Link Wray," *Guitar Player,* November 1993, p. 19.

18. Lee Hazlewood quoted in Rob Finnis and John P. Dixon, *Duane Eddy* (compan-

ion book to Bear Family Records CD set *Duane Eddy: Twangin' from Phoenix to L.A.*, 1994), p. 8.

19. Ibid., p. 22.

20. Ibid., p. 33.

21. Mike Dorrough, telephone interview by author, 19 August 1996.

22. McDermott, *Jimi Hendrix Sessions*, ix; Harry Shapiro and Caesar Glebbeek, *Jimi Hendrix: Electric Gypsy* (New York: St. Martin's Press, 1990), p. 136.

23. Shapiro and Glebbeek, *Jimi Hendrix*, pp. 215, 276.

Further Reading

Chanan, Michael. *Repeated Takes: A Short History of Recording and Its Effects on Music.* London: Verso, 1995.

Millard, André. *America on Record: A History of Recorded Sound.* London: Cambridge University Press, 1995.

6

playing with power

technology, modernity, and the electric guitar

André Millard

In the 1950s the manufacturers of electric guitars, like many other American businesses, liked to depict themselves as innovators who were constantly perfecting their product by applying the latest technologies. Americans thought of the postwar years as a new era, a time to look to the future and imagine a modern world made safer and more convenient by the fruits of advanced technologies. Intent on building their reputation as high-tech practitioners, these companies created vast numbers of innovations, both real and imagined, to help sell guitars in what was rapidly becoming a crowded market. The development of electronics technology, which produced many useful innovations applicable to guitars, ensured that the guitar manufacturers could not stand still after the first solidbody electrics appeared. The move from vacuum tube to solid state technology, for example, created opportunities to improve old products (including amplifiers) and create new ones, such as effects boxes for electric guitarists.

All the large manufacturers of musical instruments publicly referred to their R&D departments, whether these were well established and staffed by professionals (as at Gibson) or consisted of just the founder of the company and a few of his cronies. The R&D of guitar manufacturers had more than technical problems to resolve; their innovations had to be developed within the context of the popular music of the times and what sounds were fashionable with musicians.

The guitar player and part-time inventor Jimmie Webster was employed by the Gretsch company to develop guitar technology. He came up with several interesting ideas, such as the Gretsch "project-o-sonic" stereophonic bi-aural sound system. Employing split pickups for the three high and the three low strings, with controls on the guitar balancing the sound in each channel, this system was probably inspired by the success of stereo sound in home audio. It was touted as "the biggest revolution in guitar engineering since electrification."[1] Unfortunately for Gretsch, it came to nothing, but this did not stop the company or any of its competitors from thinking up new gadgets to add to their guitars to give them the appearance of modernity and superior technology, such as Gibson's Vari-tone control system.

In the opinion of Gibson's Ted McCarty the most important innovation in guitar technology was the advanced alnico magnets made from an alloy of several metals. These made pickups more responsive or "active." Leo Fender agreed with this view, but he considered that he had gone as far as he could with the pickups and devoted most of his time to improving the electronics of the amplifier and thinking up new tone circuits.[2] Whether the direction of research was aimed at the pickup or the amplifier, one critical problem was reducing the electronic noise generated during amplification—the humming, buzzing distortion that produces a jerk reaction toward the volume control if we hear it coming out of a radio or home stereo speaker.

Chet Atkins was not the only guitarist annoyed by the humming noise that came from the amp when he got too close to it. The problem was the DeArmond pickups that picked up noise from other electrical appliances nearby. Atkins had the ear of the guitar makers, especially the Gretsch company, which enjoyed his endorsement of its electric guitars. Ray Butts designed a set of pickups that did not hum and Gretsch adopted them. Gibson's laboratory had also been working on the same problem. Seth Lover invented a pickup in 1955 which had two coils (instead of one) wired in series but out of phase with one another so that when one pickup generated hum, the other canceled it out. The two-coil "humbucker" pickup was fitted to most Gibson guitars from the mid 1950s. They sounded much different from the single-coil type, having less treble end and a fatter, darker sound. The humbucker pickup was the most widely adopted innovation in electric guitars after the introduction of solid bodies.

Although the guitar manufacturers continued to develop and tout innovations in the pickup-amplifier system, none of them made much of an impact. During the 1960s each "new triumph in engineering and structural craftsmanship" of electric guitars came and went without greatly affecting the guitar market or the sound of popular music.[3]

Guitar manufacturers did not rely on technology alone to sell their guitars. Gretsch created "the most beautiful guitar in the world" in 1954 with the model 6136 White Falcon. This was a showpiece designed for trade shows, and no expense was spared in its manufacture or marketing. Described as the "ultimate" guitar, it featured metal parts plated with 24-carat gold and an iridescent white finish. It was designed to attract attention and make a statement about the style of the company's guitars and the musicians who played them. Even though it incorporated no new technology, the White Falcon was marketed as the guitar of the future.[4]

Many of the innovations applied to electric guitars were cosmetic, intended primarily to alter the look of the instrument. The acceptance of the solidbody guitar gave designers a clean slate to work with, for there was no longer any need for a sound box for amplification. They were free to draw any shape for the body of the instrument. The unusual lines of the Stratocaster had broken with the tradition of the Spanish guitar's shape and opened the way for even more unusual designs, most of which consciously aimed to give the impression of technological sophistication. The electric guitar was fast becoming a symbol of modernity for the generation born after World War II. Its literal connection to electricity made it an example of the newfound powers of electronics technology, and its association with the new music of rock 'n' roll gave it the cachet of youthful rebellion.

During the 1950s manufacturers of cars and appliances had recognized the marketing potential of projecting a sense of newness in their designs. Modern, streamlined designs for consumer and industrial products had emerged in the 1930s, but it was not until the end of World War II, and the beginning a new global era of peace and prosperity, that customers had the money to buy the idea of progress and improvement. Every kind of American manufacturer, from the big automobile companies to the makers of record players, invoked the wonderful technological advances of the war and claimed to apply them to their products. "New and Improved" was the great marketing tool of the 1950s and 1960s.

The electric guitar appeared at a time when interest and confidence in technology were at their height. The United States was looking to the heavens and embarking on an expensive, well-publicized project to explore space and get to the moon before the Russians. Both the manufacturers and the players of the guitar exploited this trend. Many of the new instrumental groups that formed around electric guitars adopted names like "the Astronauts" and wrote songs named after satellites: "Telstar" and "Sputnik," for example. Space travel in all its forms, from NASA to the Jetsons, was intriguing the American public, and any design that hinted at the technology of space travel was sure to get attention. In later years the Fender company promoted a guitar as a "space machine," for this was how it was positioned in the 1950s and 1960s.[5]

One new guitar that definitely implied modernity was the Fender Jazzmaster guitar introduced in 1957. Its body had an offset waist, which gave the guitar an unusual, futuristic silhouette. Ironically, this model was aimed not at the teenaged amateur players but rather at professional jazz guitarists (such as Les Paul). Jazz had more prestige among musicians than rock 'n' roll, and the Jazzmaster was designed to prove that solidbody guitars were as good as the traditional Gibson, Guild, and Gretsch archtops. Leo Fender changed the pickups to get a richer, more mellow sound, using wide coils like those in steel guitars to produce a broader, deeper tone. Yet the sound of the pickups and the care that Fender engineers had taken with revamping the electronics turned out to be unimportant in its commercial appeal; its modern look and high-tech controls appealed to youthful guitar buyers.

For the older, established guitar manufacturers the rise of rock 'n' roll and the solidbody electric posed a dilemma: ignore the trend and decline to make something so simple and basic (the initial response of companies like Gretsch and Guild), or forget their reputation for fine, carved-body guitars and join the solidbody camp. Should they divorce themselves from the jazz players and country pickers who had helped market their products and instead court scruffy rock 'n' rollers and amateur players? Should they stay aloof from a music that often had an unsavory image and stick with the musical genres that implied musicianship and virtuosity?

Some companies hedged their bets with innovative combinations of hollow and solid bodies and marketing strategies that aimed at both ends of the market. The Rickenbacker company, under the new leadership of F. C. Hall, quickly developed several electric guitars, beginning with the Combo line in 1954. These guitars had carved tops and a semisolid body made by scooping out the back and covering it with a plate. Gretsch also manufactured a semi-solidbody guitar. Gibson produced a thin hollowbody guitar with a solid block of wood under the pickups, tailpiece, and bridge. The ES-335 (1958) became a classic model in this series of semisolid electric guitars.

In 1957 Rickenbacker introduced a new line of electrics with a distinctive shape that still identifies the marque—the cutaways were sweeping crescent-shaped curves. A year later the top cutaway was given prominence to produce what was called "the cresting wave" shape. Crescent-shaped F holes on the hollow bodies gave the guitar a distinctive space age look, which helped differentiate it and make it easier to recognize a Rickenbacker guitar being played far away on a stage—an important consideration in the marketing of electric guitars. The question about joining or ignoring the rock bandwagon was resolved when the

Beatles toured the United States in 1964. John Lennon played a Rickenbacker and George Harrison sported a Gretsch. The demand for these models became insatiable.

Gibson's flagging sales in the 1950s had pushed the company into the solid-body camp, and it looked to that product to redefine its image as well as increase sales. The president, Ted McCarty, (who was also an engineer and designer) decided "to do something really different, something radical to knock everyone out and show them that Gibson was more *modern* than all the rest."[6] McCarty and the company's designers came up with a line of three outlandish new guitars: the aptly named Moderne, Explorer, and Flying V.

The latter was a revolutionary new look for a musical instrument, stripped down to all but the essentials and reflecting some of the totally streamlined shapes of the most advanced airplanes and spacecraft (see plate 6). McCarty chose the unusual for all aspects of the design, including rare African korina wood for the body. Called "the ultimate expression of the electric guitar's phallic imagery," the Flying V was not intended as a mass-produced guitar (only a hundred were made), but more as a promotional device for the company.[7] Most of them ended up in dealers' windows rather than in the hands of guitarists. Gibson's accountants must have considered the whole thing a complete waste of money, but they had no way of knowing how guitarists' taste would evolve over the next twenty years.

In 1963 Gibson came out with the Firebird, a toned-down version of the Explorer. Legal objections by Fender to the offset body (that looked like its Jazzmaster and Jaguar) forced Gibson to change the shape of the model: guitar manufactures were becoming more sensitive to the commercial value of design, and many claimed patent protection for the look of their products.

In addition to playing around with the shape of the bodies, designers also sought out new colors for electric guitars. The first solidbody electrics came in natural wood finishes, such as the Stratocaster's brown sunburst effect, which mirrored traditional guitar finishes, but soon more radical colors were offered, such as red or white, which represented a major change from the traditions established by a century of acoustic guitar manufacture. In an effort to differentiate their products, companies began to look at the finishes developed by other industries, especially the metallic paint used on automobile bodies. Working with the Dupont paint concern, Fender offered custom colors as early as 1956, and this led to colors like "Inca silver metallic" and the ever popular "surf green." Even the conservative Gretsch company took the plunge into new colors by adopting the metallic finishes used for drum sets, such as "champagne sparkle," for their hollowbody electric guitars. The introduction of "Cadillac green" and two-tone green guitars caused quite a stir. Duke Kramer of Gretsch said it was "unheard of" to have such colors, but things were changing so quickly that bright, unusual colors soon became the norm.[8]

amateur players and the 1960s guitar boom

The new consumers of electric guitars were not solely influenced by appearance and suggestions of modern technology. Amateur players were strongly influenced by who was playing what model. Manufacturers of musical instruments had always sought endorsements from well-known musicians, but the celebrity endorsement had special potency in the world of electric guitars, where the lure of rock 'n' roll stardom was irresistible. Electric guitar companies discovered quickly that the best advertising was to get their product into the hands of a star player.

The number of amateur players buying electric guitars increased exponentially in the 1960s as teenagers rushed to form guitar bands and the prices of instruments dropped. Mass production methods successfully applied to guitar manufacture lowered the price so that amateurs could afford them. There were numerous solidbody guitars available for less than $100; for example, the Danelectro guitar cost around $50 in 1956. Simple and functional, these instruments were made out of hardboard (Masonite) with a single pickup built into a lipstick case. Many rock 'n' rollers got a Danelectro Silvertone model C from Sears as their first guitar.

The introduction of electric bass guitars also played a role in the boom of guitar bands, because it made it possible to have an all-electric guitar group in front of the drums. These *garage bands*—a term that originally described a group of amateur players practicing in garages—were a throwback to an earlier period in rock 'n' roll, emphasizing its essentially egalitarian beginnings and the musical values of directness and enthusiasm. Michael Hicks has examined the taxonomy of the term and found a broad range of professional and amateur music associated with it. He also argues convincingly that garage as a movement provided the inspiration for other genres like heavy metal, punk, and new wave.[9]

In the amateur bands there was usually a division of labor among the guitarists—one picked out the melody with single notes while another played chords behind him, establishing the differentiation between lead and rhythm guitarists and often between leader and follower in the hierarchy of bands. They kept the music raw and simple and often did away with the vocals altogether, but as Hicks has pointed out, they created a musical culture based on the electric guitar. Up to this point, the harmonic thinking of the guitar was based on the piano keyboard, but the natural properties of the electric guitar led to a different approach to harmonies. After tuning their guitars in fourths, electric guitarists found chord progressions built around chord roots based on fourths quite familiar.[10]

The garage movement was stimulated by the great success of guitarists like Duane Eddy and his "million dollar twang." There were numerous other guitar bands with big-selling instrumental records, including several blues players like Freddie King, whose "Hideaway" was widely copied. Instrumental, surf, and hot rod music were the first expression of rock 'n' roll from the white suburbs—the sound of thousands of new groups, whose names usually began with *the:* the Ventures, the Torquays, the Surfaris, the Fireballs, the Trashmen, the Blazers, and the Beach Boys. They came from the hinterland—the West Coast, the Southwest, the Midwest, and the Northwest—well away from the musical centers of the Northeast and the musical influences of the Deep South. They were managed by parents or acquaintances and recorded in independent or home studios, and their records were released by tiny entrepreneurial labels. They still had their roots in the R&B and rockabilly that they heard on their radios and record players, as did their songs with their denatured Chuck Berry riffs, but they were coming up with a guitar sound that better reflected who they were and what they could do with their amplified instruments. They sang about cars, surfing, and girls against a solid rhythmic background of strummed chords, reverberation, and continually wavering vibrato. While maintaining the old traditions of teenage amateurism and rags-to-riches dreams, the garage bands nevertheless broke new ground in discovering dramatic electronic sounds in the vast, untapped potential of the electric guitar and its amplifying system.

The distinctive wavering slide down the low E string was the typical beginning of many surf songs, as was the overpowering reverberation that gave this music an unworldly, eerie feel. It was perhaps these metallic wavering sounds, so reminiscent of Hawaiian music, that made it easy for the listener to conjure images of the rolling waves of the ocean, but it was also a modern, machine-made sound that could not be traced back to the acoustic guitar. The constant use of the tremolo arm, which guitarists called the whammy bar, is another defining characteristic of this sound in which they raised and lowered the pitch of the notes they played (which is the vibrato rather than the tremolo effect). Guitarists switched to the bridge pickup (which had the most treble) and played heavy strings very close to the bridge to get a hard, metallic tone. They got the reverberation effect from circuits built into their amplifiers, which activated tiny springs that vibrated with the signal.

All these electronic effects went into the surf sound, which is as recognizable today as it was in the 1960s. At the time much of the attention was on the vocals—youthful, naive, and eternally optimistic—but forty years later it is the metallic sound of the surf guitars, shimmering and thundering like the waves, that makes the impression. From *Medium Cool* in 1969 to Quentin Tarrantino's *Pulp Fiction* in 1994, film directors have used this sound to give an edgy, menacing dimension to their images. Isolated by time from the 1960s and stripped bare of the surf motif in the vocals, this guitar music still sounds modern. In the television show *Third Rock from the Sun* surf guitars are very effectively mated with images of planets in the outer solar system. While instrument manufacturers invoked modernity in their designs, players did it by tweaking the switches and knobs on their equipment.

The futuristic shape of electric guitars certainly helped give the impression of a new music from outer space. The Fender Jazzmaster perfectly if unwittingly fitted this bill, because its unusual shape appealed to teenage musicians who were seeking something new and unusual. It became prominent in the garage bands. Much of its popularity was a consequence of its adoption by the Ventures, one of the most successful instrumental groups of the 1960s. Their "Walk Don't Run" (1960) was the typical catchy, guitar-driven pop song with an easily remembered melody played by the lead guitarist with the rhythm of barre chords behind it. Clean, bright, and tremolo enhanced, its simplicity encouraged teenagers all over the world to pick up an electric guitar. It entered the British pop charts just months after it had reached the *Billboard* Top Ten in the United States. The Ventures enjoyed a string of hits in the 1960s and thereby helped popularize fuzz tone, whammy bars, and Fender guitars, which were always prominent in the images of the band. The picture on their LP shows four clean-cut youths with matching outfits and big smiles in a pose reminiscent of the lighthearted attitude taken by the first jazz players in the 1920s. The Ventures appear as respectable as any bunch of kids from the suburbs.

The other favorite of instrumental bands was the Mosrite guitar. Semie Moseley was a tinkerer and custom guitar maker who had worked for Rickenbacker under Paul Barth. He had set up limited production in a barn in Bakersfield, California, when he met Nokie Edwards of the Ventures and went into partnership with him to produce a new guitar. Moseley had designed plenty of guitars in his time, but like everybody else he started with the shape of the Stratocaster. The Ventures model guitar was an upside-down Strat (meaning that the lower cutaway horn is longer) that Moseley mass produced with financing from the Ventures management. He was soon selling hundreds of guitars a month, most of them to garage bands trying to sound like the Ventures.

amplification

The sound of the garage bands was dependent on their amplifiers. The same advances in electronics that were driving innovation in electric guitars were equally important in amplifiers and effects. More so, in fact, because the 1960s marked the identification of loudness and distortion as the critical ingredients in the sound of the solidbody guitar. Although the amplifier is usually considered to be an adjunct to the instrument, its role in creating rock 'n' roll's unique voice reached the point where some amplifier manufacturers, like Marshall, could claim that their products were responsible for "The Sound of Rock." During the 1960s manufacturers of amplifiers could hardly ignore the input from musicians who were clamoring for more power. This reflected the current aesthetic in popular music that louder was better, but it was mainly the result of the increasing size of the audience; bands were playing bigger venues and needed more volume. The increase in amplification power is a clear indicator of the electric guitar's growing importance in popular music.

The amplifiers available to power the first solidbody electric guitars were small and underpowered. They came in a large case, which held the single speaker and usually had only two inputs (one for a microphone and one for an instrument—often an accordion!), along with volume and sometimes tone controls. Designers of guitar amplifiers had to use existing vacuum tubes, which were manufactured under license from RCA, and usually followed the circuits published by RCA in their tube manuals. The 6V6 power tubes available in the 1940s limited amplifier output to around twenty watts using the conventional single tube. Fifteen watts of output was considered to be more than enough power for electric guitarists. Gibson used a twelve-watt amp—no bigger than a large briefcase—to power its first electric guitars.

Even after the successful introduction of the Telecaster, Leo Fender was known more for his amplifiers than for his guitars. His first models, introduced in the 1940s, were small amplifiers in wood- or leatherette-covered boxes with outputs from six to fourteen watts (see plate 7). He built larger and more powerful units for the professional players, such as his Dual Professional of 1946—one of the first twin speaker models—and the Professional, which came with a fifteen-inch speaker and forty watts of power. This was a big jump from the ten to fifteen watts that was the norm for musicians' amplifiers at this time.

Increasing the power of the amplifier put more stress on the loudspeaker. Small companies like Jensen and JBL had benefited from wartime research in electronics and acoustics and were exploiting new materials like alnico for their

The Rolling Stones performing at Hyde Park. Banks of speakers and amplifiers were now needed to reach the huge crowds at rock concerts. You enjoyed the show within a deafening envelope of sound. Photo by Barrie Wentzell.

loudspeaker magnets. The JBL D130 speaker was used in several Fender amps, such as the Twin series, because of its innovative voice coil design (to reproduce low frequencies) and its rugged construction, which enabled it to stand up to overloading by players. Some guitarists were playing so loudly and aggressively that they forced the cones out of the speakers, and bass players regularly blew out their speakers with loud, overloaded low frequencies. Fender reasoned that the single fifteen-inch speaker in his bass amp could not stand up this kind of playing and replaced it with four ten-inch speakers—the celebrated Jensen P10R. They not only resisted blowing out but also created a warmer, choruslike effect. These were the voice of Fender's Bassman amplifier that gave the garage bands' sound some substance or "bottom." The amp also had piercing treble and broke up into a pleasing distortion when overloaded, which made it a favorite of blues players.

The cooperation of speaker companies was an important factor in the development of guitar amplifiers, which had to stand up to unusual demands and abuse. Fortunately, many were located in southern California, which was becoming a center for guitar and amplifier manufacture. The interaction of speaker manufacturers and designers of amplifiers, who were in turn using players to evaluate and field test their prototypes, helped shape the evolution of guitar amplifiers.

The rated power of amplifiers grew steadily in the 1950s. The signal from the guitar's pickups is extremely weak (about .0025 volts) and has to be increased before it can be manipulated within the amplifier. This is the job of the preamplifier, which increases the signal to about one volt in the "gain" circuit. Next come the processing circuits, where the signal is shaped by altering bass and treble and where reverberation and other effects can be added. As effects became more popular, manufacturers added additional amplifying channels to create numerous gain and processing circuits in their amplifiers. After processing, the signal enters the output circuit, where power tubes increase the voltage to about eight volts and create the wattage to power large speakers. More powerful power tubes were now available—especially the RCA 6L6—which increased output. Designers employed more than one single output tube to get more power: the more power tubes, the higher the output of the amplifier.

Fender's 1958 Twin amplifier had four power tubes and twice the wattage of the Bassman. The guitarist Dick Dale, who earned the title King of the Surf Guitarists, worked closely with Fender to develop higher-output amplifiers. He field tested the Dual Showman amplifier, which was a "piggy back" amp in which the electronics were packaged in a separate box from the speaker enclosures. With more large speakers in use, amplifier cabinets were naturally getting larger and heavier. Bands that needed even more volume were now running several speaker cabinets from one master amplifier.

Musicians' interaction with their amplifiers provided an initiation into the world of electronic technology. They began to tinker with the amps just as they had messed about with the pickups on their guitars to get a different sound. Some discovered new sounds in equipment that was breaking down; others deliberately did the breaking themselves. Rock musicians were highly discerning in evaluating the tone of their amps, analyzing the sound and using terms like *crunchy, swirly, smoky, booming, fat, flubby, doinky, dirty, greasy,* and *country clean* to describe it.

The sound effect that rock 'n' roll and rockabilly guitarists found most attrac tive in the underamplified 1950s was echo, a sparse electric echo that gave depth and a certain ominous drama to the sound. "Heartbreak Hotel," for instance, has one that makes the plaintive lyrics come to life and gives the listener a sense of the anguish and solitude suffered by the singer. Echo was created by electronically duplicating a note right after it was struck, and the amount of delay (in milliseconds) between original and electronic copy determined the characteristics of the echo. Early rock 'n' rollers used a single "slap back" echo that came right after the actual note. Although this is often referred to as the Sun Studio echo—after Sam Phillips's tape delay procedure—it was conjured up in many recording studios. Effects like echo were used to personalize the sound and add unique character and sonorities, which helped create a identity for the player or recording studio.

Recording engineers were indeed crafting the sound of popular music with electronic effects, but the technology could be applied to other parts of the system, such as the amplifier. Running parallel to innovations in recording studios was the development of circuits that manipulated the signal produced by the guitar. Special channels in amplifiers or stand-alone effects boxes promised an easier way of crafting the sound than the recording studio, which sold its time and expertise dearly and was often out of bounds for the amateur guitarist. Many of the effects created by recording engineers, such as echo, could be inserted in the signal chain or the amplifier. The Vox company marketed its line of effects devices with the claim that they placed "the facilities of a recording studio at your fingertips," which indeed they did. To reproduce the commercial sound of rock 'n' roll in the late 1960s, one either had to experiment in a recording studio or dial it up on the amplifier or effects box.

The freelance designers and small businessmen who produced effects boxes (and later pedals) were cut from the same cloth as the originators of the electric guitar: amateur inventors, tinkerers, and entrepreneurs in the radio, television, and amplifier businesses. Ray Butts, a radio repairman and owner of a music store in Cairo, Illinois, was one of the first to incorporate echo circuits in amplifiers. He found existing amplifiers inefficient and thus built his own and added a tape echo device to his EchoSonic amplifier, which was adopted by several influential guitarists. One was Scotty Moore, who used it on many of Elvis Presley's Sun and early RCA recordings. Soon several companies were marketing stand-alone echo devices. The most popular was the Echoplex unit, which was based on Butts's design. The Echoplex became a fixture in studios creating rock 'n' roll: the secret ingredient to the distinctive sound of many of the genre's most important recordings.

Another popular effect was reverberation. This increases the sustain of a sound and makes it appear to be reverberating within the confines of an enclosed space. Whereas echo produces several discreet sounds, reverberation produces a continuous sustaining of pitch. Recordings carried out in large, reverberating spaces like concert halls or stairwells have that rich, ambient quality of the original sound plus its reflections. When Dick Dale asked Fender to create a device to increase the sustain of his voice (because it lacked a natural vibrato), Fender investigated the electro-mechanical reverberation employed by organ manufacturers and licensed the idea from the Hammond Organ Company. Naturally, Dick Dale played around with the new unit and found that when plugged in to the pickup (rather than the microphone) circuit on the amplifier, it produced an appealing reverberating guitar sound. This was incorporated into many of his songs, including "Miserlou" (1962). The Fender 6G-15 Reverb unit was introduced in 1961 and produced an effect that some described as a wet, splashy sound. It can be heard on recordings such as the Chantay's "Pipeline" (1963).

The popularity of the surf sound encouraged designers to add reverberation channels to amplifiers. Some also incorporated the tremolo effect. By 1963 the acclaimed Fender Twin Reverb amplifier had a rated output of eighty-five watts, more than double the power of Fender's first professional amplifiers fifteen years earlier. Musicians were getting more than power from this new generation of amplifiers, because they could now control the tone, intensity, and duration of several built-in effects.

The development of amplifier and effects technologies was not confined to the United States. Europe also had trained people with skills gained in the wartime development of communications, radar, and computing. They saw ways of improving guitar amplifiers and recognized a business opportunity in the rise of American popular music that British guitarists were now set on copying. American guitars and amplifiers were not widely available outside the United States, and this encouraged a business of importing them or, better still, copying them. With the same entrepreneurial spirit that characterized many of the electric guitar manufacturers, Europeans set up their own small businesses to market their innovations.

Dave Reeves worked as an electrical wirer for the English Sound City company before branching out into making his own amplifiers. His Hiwatt amps were the real beginning of rock-guitar amplification in the United Kingdom. Dick Denny was an electrical engineer (and guitar player) who befriended Tom Jennings during the war. Jennings owned a music shop in Dartford (near London) and was most impressed when Denny showed him an amplifier he had built. They formed a company and the first Vox AC15 (indicating an output of fifteen watts) amps were produced in 1957, followed by the AC30 in 1959. In the hands of the Beatles and the Rolling Stones, the Vox AC30 became prominent in 1960s rock music.

Jim Marshall owned a drum store in London. A musician who had received some technical training during the war, Marshall imported American equipment to sell to British bands. He stocked guitars and amplifiers and had no problems selling what was then very desirable and expensive equipment. He recognized that while electric guitars were too much of a challenge, he could easily make his own amplifiers based on the Fender products he had disassembled in his shop. In 1962 he collaborated with Ken Brown (one of his customers) and teenage electronics wizard Dudley Craven (who worked for the EMI company) on the first Marshall amplifier, which they put together in his repair shop using 5881 power tubes (instead of the 6L6) and following the layout of the Bassman circuit. Legend has it that Marshall could not manufacture amps quickly enough to satisfy his customers. A year after he built his first amp, he had set up a separate factory to manufacture them; by 1964 he was in a larger factory producing about thirty amplifiers a week.

The European amplifiers might have been Fender copies, but they were not exact copies, because they were made with European components. Instead of the hard-to-find 6L6 power tubes of the American product, Marshall used British-made KT66s and Vox used EL84s, which were more easily acquired in England. European vacuum tubes were made under license from U.S. companies, but musicians agreed that they sounded warmer and had more harmonic distortion than their American counterparts. This made their amps sound a little different from the American amplifiers and helped distinguish them. Vox was renowned for its warm midtones and Marshall for the fat and aggressive sound at the bottom end. By the mid 1960s these brand names stood on their own merits and entered the American market by way of touring English bands.

the search for controllable distortion

One major consequence of the growing loudness of rock 'n' roll was the increased presence of distortion. One of the reasons that the early R&B records sounded so different and exciting was because of the fuzzy, humming tone caused by distortion in overloaded amplifier circuits. (This muddy sound was enhanced if the record was played on a small, lo-fi portable player.) While the makers of amplifiers and the designers of pickups treated distortion as a major technical problem, many guitar players welcomed it as they sought new and more expressive sounds.

Turning up all the volume controls to full produced distortion. Overloading the pickup-amplifier system—playing in the red warning zone of the meters—was something that musicians were not supposed to do, but as the volume of signals overloaded the preamplifier vacuum tubes, a new catalog of electronic noise emerged: distorted, reverberating machine-made sounds that turned out to have considerable musicality and harmonic potential. The first people to use distortion creatively were not engineers in the recording studio but African American players who exploited the inadequacies of their amplifiers to add an electronic edge to the blues. Legendary guitarists like Clarence "Gatemouth" Brown and Guitar Slim employed feedback and distortion in their music, and they also played louder than many of their peers. Most amateur musicians came to distortion by inadvertently overloading their amplifiers or plugging in several guitars into one amp, or both. This was especially the case when they were using low-wattage amplifiers in large spaces. Unable to eliminate distortion, some players attempted to control it: "I decided to use it rather than fight it" was the approach of English guitarist Jeff Beck.[11]

As the sound of rock 'n' roll was becoming colored by electronic tones, more amateur musicians heard these new distorted sounds on record and wanted to make them. Here was an opportunity for the independent inventor and small businessman. Ivor Arbiter, who developed the Fuzz Face distortion pedal popularized by Jimi Hendrix, built up Arbiter Electronics on the success of his effects boxes. Charlie Watkins established Watkins Electric Music to manufacture his own amplifier designs and his wildly popular Copicat tape echo box. Many of these British entrepreneurs already owned, or soon acquired, their own retail outlets. These small-scale concerns received a mighty push if a well-known musician started to use their equipment.

Companies manufacturing amplifiers saw effects boxes as a valuable sideline that incorporated the technology already devised for amplification. The English Vox company branched out into electronic organs and electric guitars as well as effects devices such as the Vibravox vibrato unit and the Vox Echo. Its engineers carefully examined all the latest equipment coming from the United States in their development of the Vox Tone Bender fuzz box. Vox might have been considered a follower in the effects field until celebrity guitarists like Jeff Beck and Keith Richards adopted their boxes.

Several electric guitar manufacturers diversified into effects to take advantage of their name recognition and marketing organizations. One of the most suc-

Clarence "Gatemouth" Brown. The highly influential Texas blues player was one of the African American pioneers of playing loud and hard. Courtesy of Jonas Bernholm Collection, Division of Cultural History, National Museum of American History, Behring Center, Smithsonian Institution.

cessful fuzz boxes was the Maestro Fuzz-Tone, which was introduced in 1963 by a company owned by Gibson. In addition to volume it featured an "attack" control that went from "faint" to "biting." Semie Moseley had his electronics technician Eddie Sanner develop their own fuzz effect, the Fuzzrite, which was introduced in 1966 first with germanium transistors, then with silicon transistors.

Ironically, the circuit in the effects box that actually duplicated the sounds of overloaded vacuum tube amps did so with the much more advanced transistors. The transistor had begun to replace vacuum tubes in many applications at this time, because it was smaller, more efficient, and less prone to heat up, fail, or break. It was touted as the future of electronics; but there was one small enclave in the world of amplified sound that insisted on staying in the past. Musicians preferred the sound of vacuum tube amplification, arguing that it was warmer and sounded more organic. They used words like *brittle, sharp,* and *clinical* to describe the sound of transistor amplification and lamented the lack of distortion. Consequently, the amplifier manufacturers' attempt to move up to the next stage of electronic technology was not a commercial success. Whole lines of modern-looking amps with the latest transistors failed to replace the old standards, which were now eagerly sought out by professional musicians.

Making amplifiers was now as big a business as manufacturing electric guitars, and many of the big names in electric guitars moved into amplifiers: Gibson, Fender, Guild, Gretsch, and Rickenbacker. There was also a second tier of smaller makers who came from the audio or musical instrument businesses. Ampeg was formed by Charles Everett Hull to make pickups but soon moved into amplifiers. Tinkerers in retail outlets and guitar repair workshops also saw a profitable new area for innovation. The boom in home stereo—higher-fidelity long-playing records, two-channel amplifiers, and lower noise-to-signal ratios—had brought more engineers and manufacturers into the business, and many of the leading designers of guitar amplifiers, such as Leo Fender and Everett Hull, kept a close eye on what Fisher, Marantz, and MacIntosh were developing for the audiophile. The latter's objective was plain and simple: pure, distortionless sound. The former were divided: some, like Hull, kept to the principle that distortion was a sign that the amplifier was not well designed; others, like Fender, heard the musicians who excitedly talked about the new sounds they could get from overloaded amplifiers.

As rock music entered the psychedelic era of the midsixties, the distorted, fuzzy sound became the fashion. Keith Richards' guitar playing on "Satisfaction" (1965) sold a lot of Maestro Fuzz-Tone boxes, but equally influential were the musicians who used this effect for television and movie soundtracks, such as

Davie Allan, who created dramatic soundtracks for films like *The Wild Angels,* which were aimed at teenagers. A fuzz tone and an echo box were now considered standard equipment for the rock guitarist, and the market was saturated with boxes with names like Distortion Booster, Pep Box, and Astrotone.

This obsession with distortion was a far cry from the stated goal of amplifier manufacturers: "high fidelity, capable of producing considerable power without distortion."[12] In the 1960s many musicians had a different agenda, and they could make choices by altering the controls on their amps or adding effects boxes. These customizing options were often at odds with the technological strategies of the manufacturers and prevailing notions about the operation of equipment. But if you relied on the instruction manual, you would not get the right sound. The inventor Roger Mayer said that creating effects devices "becomes an exercise in knowing exactly what to do wrong."[13]

The people who bought guitars and amplifiers did not always follow the manufacturers' recommendations, nor did they act like rational buyers. The strategy of technological innovation did not always work, nor did the modernity implied in equipment design. Rock guitarists were ambivalent about modernity, picking and choosing between antique classics and totally new products. The acceptance of these innovations, like the music, was more likely to be influenced by fashion and personality—especially the cult of the rock star—than by the best efforts of engineers and designers.

Sources

Advertisements from *Guitar Player*

Doyle, Michael. *The History of Marshall.* Milwaukee: Hal Leonard, 1993.

Flieger, Ritchie. *Amps! The Other Half of Rock'n'Roll.* Milwaukee: Hall Leonard, 1993.

Smith, Richard. *Rickenbacker: The History of Rickenbacker Guitars.* Fullerton, Calif.: Centerstream, 1987.

Thompson, Art. *Stompbox.* Milwaukee: Hall Leonard, 1992.

Wheeler, Tom. *American Guitars: An Illustrated History.* Rev. ed. New York, HarperCollins, 1992.

Notes

1. Gretsch advertisements reproduced in Jay Scott, *The Guitars of Fred Gretsch* (Fullerton, Calif.: Centerstream, 1992), p. 47.

2. Ted McCarty, interview at National Museum of American History, Washington,

D.C., 15 November 1996; Richard Smith, *Fender: The Sound Heard around the World* (Fullerton, Calif.: Garfish, 1995), p. 150.

3. Advertisement for Rickenbacker twelve-string guitar, *Guitar Player,* July 1967.

4. Scott, *Guitars of Gretsch,* pp. 104–14.

5. Advertisement for Fender guitars, *Guitar Player,* March 1973.

6. McCarty quote from interview, emphasis added.

7. Paul Trynka, *The Electric Guitar: An Illustrated History* (San Francisco: Chronicle, 1993), p. 95.

8. Kramer quote from interview in "Electric Guitar Video Documentation," 9–16 November 1996, National Museum of American History, Washington, D.C.

9. Hicks, *Sixties Rock,* pp. viii, 25–27.

10. Ibid., p. 35.

11. Jeff Beck, interview in *Guitar Player,* November 1973, p. 22.

12. Fender catalog, 1956.

13. Steve Waksman, *Instruments of Desire* (Cambridge: Harvard University Press, 2000), pp. 183–84.

Further Reading

Hicks, Michael. *Sixties Rock: Garage, Psychedelic, and Other Satisfactions.* Urbana: University of Illinois Press, 1999.

Palmer, Robert. *Rock & Roll: An Unruly History.* New York: Harmony Books, 1995.

7 the guitar hero

André Millard,
after the collaborative paper
with Rebecca McSwain

What is a guitar hero? A man, a solo artist, a virtuoso, certainly, but more than that: someone whose willingness to experiment leads the instrument to new dimensions, someone who is widely imitated, and someone whose very life seems to define the instrument in a new way. The idea of a guitar hero is no longer confined to musicians; it is now part of our popular culture, and the term *guitar hero* is used (untranslated) in other cultures.

Who is a guitar hero? Many of us who grew up in the 1960s would simply answer "Eric Clapton." Others, coming of age in the 1970s, might say "Carlos Santana" or "Peter Frampton." Still others, youngsters in the 1980s, would respond, "Eddie Van Halen." Virtually all guitar players, regardless of age, sex, or national origin, would say "Jimi Hendrix." The list would grow as we asked more people: Peter Townshend, Les Paul, Jeff Beck, Jimmy Page, Slash, Steve Vai, Ritchie Blackmore, Eric Johnson, Yngwie Malmsteen, Joe Satriani, Stevie Ray Vaughan. Teenagers might want to include Tom Morello of Rage Against the Machine or "Munky" Shaffer and "Head" Welch of Korn.

Whatever the opinions (and disagreements) of music lovers about who should be placed in the pantheon, there is a general consensus of the qualities that a guitar hero has to have. There must be virtuosity, certainly, but skill at playing the guitar is not the only qualification. There is a heroic dimension of the guitar hero that encompasses lifestyles and values stretching back far beyond the invention of electric amplification. The hero was part of Greek mythology, but the guitar hero emerged in the 1960s when electric guitar technology merged with an English interpretation of African American blues traditions and some timeless Western ideas of masculinity.

The seeds of guitar heroism were sown by African American blues guitarists of the 1920s and 1930s. At a time when bands (black and white) used the guitar as a rhythm instrument, the bluesmen (and an occasional blueswoman) were soloists whose livelihood depended upon their skills. Virtuosity was no doubt an advantage, but it was more important to make a powerful emotional connection with the audience—these musicians were, after all, playing for tips. The blues personalized the music and focused attention on the singer, but it was a statement of the collective experience of Africans in America. This function was carried on by the African American players who interpreted the blues with electric guitars in the 1940s and 1950s and who lived the same itinerant lifestyle of the blues player. From the point of view of the young men who were discovering the blues and learning to play guitar in the 1960s, the bluesman lived a romantic life on the road not too different from that of the rock musician. The bluesman was a solitary, swaggering individual who lived by his wits and enjoyed the fruits of his playing. He was a powerful man, a hoochie coochie man, as Muddy Waters put it, both gifted and dammed by the gods, and one who had special powers—especially over women.

The myth of the bluesman was established in the songs, which usually identified the narrator as the hero. Although the blues appeared in the early twentieth century, the self-conscious stars of rock music found it remarkably appropriate. These guitar players identified with the blues masters of the past who had often lived short, fast, famous lives. "Show me a hero and I will write you a tragedy," said F. Scott Fitzgerald, and the tragedy—even (or perhaps especially) self-inflicted tragedy—became an integral part of the heroic role.

First, there were the sacrifices that the aspiring bluesman had to make to learn his art. Rock 'n' roll has its own work ethic in the paying of one's dues: enduring years of low-paying gigs and thousands of miles on the road. The Allman Brothers represented this tradition, living on the poverty line and touring relentlessly for years, but enjoying fellowship before they made it to the big time. As established stars they were still expected to remain true to their roots. It was Duane Allman who said that they don't pay rock stars to dress up funny but to play music. In the mythology of the bluesman the paying of one's dues could go beyond simple hard work and devotion to the music. Robert Johnson, the king of the Delta blues players, was reputed to have sold his soul to the devil in exchange for his great powers on the guitar. When he sang "Hellhound on My Trail" you could almost believe this story and imagine the devil coming to claim his price. Some of Eric Clapton's fans will claim that he spent many years learning how to play the guitar in locked rooms, with only drugs as his companion.

The guitar played a vital part in the myth, and there was a close physical and psychological relationship between player and instrument. Blues guitarists often referred to the guitar in terms of sexual imagery; the early bluesmen called their instruments "easy riders," which on one level refers to its being carried on the back but also means a lover—the female partner is also called an "easy rider." Robert Johnson sang "Going down to Rosedale, take my rider by my side" in "Crossroads Blues." Musicians in the era of rock 'n' roll also described their guitars as female sex objects: "She was a delicate shade of pink and was one of the most sensational pickups I've ever seen. I had to have her. I longed to let my fingers stroke and caress her." Even the young guitar heroes of the twenty-first century maintain this tradition. Asked about what guitars he uses, Kenny Wayne Shepherd put his 1961 sunburst Stratocaster at the top of the list: "That's my woman."[1] The rumors were that Jimi Hendrix never let his guitar out of his sight, even taking it to bed with him at night. Other guitar heroes reputedly asked to be buried with their guitars.

The guitar functions in the legend of rock stardom in several ways. The biographies of star players often begin with a homemade guitar, fashioned out of scraps of wood and pieces of wire—an embodiment of boyish technological enthusiasm and an indication of the powerful forces pushing the artist to create. Guitars are built, loaned out, sold, exchanged, and even given away in some cases. The informal histories of the Allman Brothers contain the story of Duane Allman giving one of his trademark Les Paul guitars to an admiring fan. (Stories like this are also common in country music—where stars' kindness to fans is an important part of the sense of community.)

The guitar also figures in the travels of the bluesman. The hero of Greek legend normally undertook a journey, and the stages of a hero's journey are essentially the same whether it is Ulysses, Andrew Carnegie, or Jimi Hendrix. First comes separation from home and the ordinary, then a journey with several challenges that must be overcome, and eventually after many adventures the return home with self-knowledge; a better, enlightened person emerges at the end of the story. The journey of the guitar player of blues legend and the heroes of rock 'n' roll all existed within the American tradition of economic advancement, but the rewards were much more likely to go to the modern players who had the education, the business sense, and the legal advice to ensure proper payment.

The guitar heroes of the 1960s enjoyed unprecedented rewards, becoming celebrities who were known for a lot more than their playing. Rock stars enjoyed great wealth, but it was not all about money; many of them were literally worshiped by their fans, for example, Eric Clapton, the most revered and imitated guitarist of his generation. His speed and virtuosity were legendary, and he quickly created a signature blues-rock sound based on humbucker pickups on Gibson guitars driving Marshall amps. Such was the idolatry of this self-effacing musician that "Clapton is God" graffiti appeared all over London.

In blues culture there is something frightening about the extreme virtuoso. He is not like other men; what has he sacrificed to become a hero? In the case of Robert Johnson, it was his everlasting soul. The idea of a hero as a virtuoso whose abilities redefine the possibilities of the instrument is not intrinsic to the blues nor to the guitar. It goes all the way back to the story of Orpheus. The gift of genius usually comes with a few drawbacks, ranging from mild antisocial behavior to bouts of insanity or even loss of life. All these characteristics are present in our modern articulation of the guitar hero.

To bring the guitar hero into existence, the myth of the bluesman merged with the technologically innovative English blues bands of the 1960s and their mass audience on both sides of the Atlantic. The blues tradition established the importance of the solo player; the electric guitar virtuosos married that tradition with technology, using all the potential of high volume, sustain, distortion, and feedback.

tools of the guitar hero

In their efforts to make the electric guitar a louder and more versatile instrument, the manufacturers also created the means to make its player a star. In the big bands of the 1930s and 1940s the guitar player was a member of the rhythm section, seated toward the rear of the band while the brass and reed players usually sat in the front, but the electric guitar player of the 1950s stood as a prominent figure in the smaller groups of R&B and rock 'n' roll bands. With a lightweight, compact solidbody instrument, the guitarist was free to move about, restricted only by the umbilical cord that connected him to the amplifier. Although the guitar player was a lot more prominent, he was still not the star of the show. He had to compete with saxophone players and pianists for the limelight. Rock 'n' roll emerged at a time when the male vocalist—such as Frank Sinatra or Hank Williams—dominated popular music. Elvis Presley held an acoustic guitar, but the significant guitar sound came from Scotty Moore in the background. Although the glamorous vocalist had always had a place in rock 'n' roll, gradually more and more guitarists shared the spotlight.

The first groups of amateur rock 'n' roll players often had a vocalist–lead guitarist who played a guitar and delivered the vocals in front of the drummer and the bass player. Players like Eddie Cochran and Buddy Holly led their bands from the front. Live appearances, photographs, and images on record covers placed the electric guitar at the very epicenter of rock 'n' roll. It was not a stationary instrument; the guitarist was expected to move around rather than just stand in front of the amplifier and just play. The guitar strap was therefore a vital addition to the instrumentation of rock 'n' roll. As the prominent zoologist Desmond Morris has pointed out, it allowed the player to lower the instrument to crotch level and use it to make symbolic gestures.

Chuck Berry and countless other African American R&B artists had developed moves with their guitars as part of their stagecraft, borrowing from those who went before them, especially the great showman T-Bone Walker. Berry perfected his duck walk in which he crouched and strutted across the stage, and other players went a little further in holding the guitar in a suggestive manner. The electric guitar became the important part of a show of masculinity, and perhaps because of the influence of the blues, it was specifically associated with black masculinity. In his book on the electric guitar Steven Waksman calls it a "technophallus"; he sees it as part of white attempts to appropriate the sexual potency of black men.

It is unlikely that players like Buddy Holly or Eddie Cochran thought of the guitar in these loaded terms, but they and their peers eagerly took part in a modern update of the minstrel show when they copied the moves of celebrated black players—like the entertainers of the nineteenth century who adopted blackface in their attempts to imitate African American song and dance. Rock 'n' rollers sometimes recreated stereotypes of African Americans as simple, emotional hedonists more in touch with their emotions and more overtly sexual than uptight Anglo-Saxons. Deciding whether the guitar was instrumental in this spectacle or just a prop depends on how heavily one invests it with significance and sexuality.

The status and signifying power of the solidbody electric guitar in popular music grew with its amplified volume. The all-guitar instrumental bands of the 1960s did away with the vocalist altogether and made the lead guitarist the voice of rock 'n' roll. Duane Eddy, Dick Dale, Link Wray, and Nokie Edwards were the models for thousands of amateurs who bought their records and copied their licks. Respected for their virtuosity and held up as examples of professional development, each had a distinctive sound, a style on stage, and an association with a particular instrument: Buddy Holly and Dick Dale were devotees of the Strato-

T-Bone Walker with a Gibson ES-150. This studio shot shows him standing still—something you would have rarely seen in his act. Courtesy of Frank Driggs Collection.

The great showman Chuck Berry doing his duck walk. Now in his seventies, he still performs his ageless songs. From the collection of Rock and Roll Hall of Fame and Museum.

caster; Duane Eddy used a Gretsch; Nokie Edwards was associated with the Jazzmaster, then the Mosrite; and Link Wray often played a Danelectro guitar.

Their followers coveted these guitars as status symbols. The top-of-the-line Gibson or Gretsch carried more weight than just a musical instrument; it was a symbol of success and a token of professional attainment. Every band proudly displayed its new guitars on stage and record covers. The first real sign of success in the competitive world of rock 'n' roll was the new guitar, and each stage in the rise of the teenage band was marked by a different instrument: a cheap Sears Silvertone for the beginners, a worn secondhand instrument purchased from another musician as the band got more gigs, and finally a brand new model with a premium name. At the summit of success was a customized model named after the player and bearing his name—this marked hero status.

The high schoolers working their way through the simple chord progressions of Chuck Berry or the Ventures practiced their moves as well as their fretting, for this was an important part of the music. They also copied clothes and hair styles—you really could not prepare for a career in rock 'n' roll without a mirror. But more than any other artifact, the electric guitar was the talisman of the creativity and acclaim that motivated thousands of young musicians. It was as if the genius was incorporated into the solid bodies and electronic circuits. Acquiring the kind of instrument used by the star was an important step in becoming more like him; this was an investment, not a fantasy, and it produced a correspondence whose authenticity could not be questioned. This process of association with virtuoso players via the make and model of guitar was behind the greatest boom in sales the manufacturing industry had ever experienced.

Guitar worship among teenagers was an international movement; it followed the global distribution of recordings and the ever widening range of tours. Films and television shows in grainy black and white provided the image to be emulated: Europe had its Elvis clones, its Ventures copies, and its Buddy Holly sound-alikes. The Shadows were the first English guitar band to emerge with a hit instrumental record, "Apache," in 1960. This song sounds like the soundtrack to one of the cowboy shows that were popular on English television at the time, with a background of beating Indian drums and a guitar twang that would have sounded more at home in Wyoming than it did in London. But this was not considered much of an anomaly in a country slavishly following every twist and turn of American popular culture.

The Shadows were one of the most successful English groups of the 1960s, and they were led by a young man as bespectacled and as earnest as Buddy Holly. Hank Marvin was the first in England to acquire one of Fender's beautiful new

Stratocasters. It was chosen sight unseen from the Fender catalog, and when it arrived, "no one touched it for a while: they just stared at it. The guitar was the most beautiful thing they had ever seen." The salmon-pink Stratocaster was given pride of place in the Shadows' stage show and on its album cover. It inspired more than one budding guitarist, as Pete Townshend remembered of one of his early band-mates: "His dream was to own a Fender Stratocaster, a pink one, like Hank's and to front a group playing Shadows material."[2]

After World War II import restrictions made American goods unavailable in the United Kingdom, so the first solidbody electrics came from European manufacturers such as Hofner. American guitars were rare and expensive, and this gave them even greater cachet: "any good American guitar looked sensational to us. We had only beat up, crummy guitars at that stage" was how a couple of musicians remembered it. Even the Phillips screws seemed modern and advanced to Europeans accustomed to just one cut on the screw head. American electric guitars represented more than Yankee ingenuity; they were icons of the rock 'n' roll universe spreading into Europe. The Fenders and Gibsons existed in the unreal and distant stratosphere of American stardom, and these unattainable guitars were "played only in our dreams."[3]

Although scholars have followed the diffusion of rock 'n' roll music into other countries, the emphasis has been on the recordings, not the hardware; yet guitars and amplifiers were just as important in the globalization of American popular culture. Nowhere was the electric guitar more revered than in England. Charlie Gillett pointed out in his history of rock music that English kids took the cult of the guitarist more seriously than their American counterparts did. The British music audience was bursting with admiration for the lead guitar players, like Buddy Holly and Eddie Cochran, and often made them bigger stars in Europe than they were in the United States. While Duane Eddy and Dick Dale were acknowledged as the pioneers of the guitar instrumental in Europe, they sold nowhere near as many records as the Shadows, who were supported by an army of amateur players.[4]

The guitar instrumentals of the 1960s were easy to learn and universally duplicated in Europe, but a few English guitarists were a different breed from the journeymen strumming their way through the Shadows' hits. Their skills with technological manipulation, their loudness, and their virtuosity set them apart from other guitarists. They became the great guitar innovators in the early 1960s, creating not only new sounds for the instrument but also a new prestige for the virtuoso player. They quickly moved from local prominence in the United Kingdom to become universally recognized guitar heroes in the world of rock.

volume and velocity

Playing the guitar was popular among teenagers in England in the 1960s because many of them were inspired by a major blues revival. This music had long been in neglect in the United States, ignored by American whites in a segregated society and forgotten by many African Americans, who preferred to put that part of their history behind them. The American folk revival had brought some attention to the acoustic country blues, but there was not nearly the respect and interest felt in other countries. Europeans heard an immediacy and power that was lacking in their own popular music; this is a good example of alienated individuals from elite groups deriving comfort from the culture of exploited and aggrieved populations. Identifying with a group whose class and ethnicity differ from one's own provides an alternative to the suffocating norms of growing up in a nice middle-class environment. European teenagers were taken by the romantic notion of the musician as outsider, and no musician was as outside the mainstream as the bluesman—lonely, broke, and black. The civil rights struggle in the Deep South was extensively and sympathetically reported in Europe, adding weight to the authenticity of African American culture. As Eric Clapton reminisced about his youthful attachment to the idea of the lone bluesman: "It was one man and one guitar against the world. . . . it was one guy who was completely alone and had no options . . . other than to sing and play to ease his pain."[5]

For all these reasons young English players usually knew more about blues giants like Robert Johnson than their American counterparts did. It was not just the music that intrigued them; the lives and legends of the blues guitarists were eagerly absorbed by young men who had never lived in Brownsville or Rosedale but had visited them in the imaginary world of recordings. Perhaps it was because of this European interest that several American record labels began to reissue classic blues albums in the 1960s. English guitarists had educated themselves on American R&B, and after they had their fill of Chuck Berry and Little Richard, they graduated to Muddy Waters and John Lee Hooker and from there went all the way back to the country blues of Robert Johnson and Blind Lemon Jefferson, which was now available on long-playing records. This was part of a search for authenticity in popular music that had led fans away from slick professionalism to the folk and blues revival, not the electric blues of the city but the acoustic blues of the country.

Bands like the Animals, the Rolling Stones, and the Yardbirds brought attention to the old blues masters by playing their songs. They gave the blues much more respect than it ever had in the United States, and these groups' growing popularity among American audiences reintroduced it to its homeland. As the blues guitarist Buddy Guy noted: "they came to this country and told the people where they got their ideas from. . . . A lot of white people never knew about our great musicians."[6] When Eric Clapton (of the band Cream) played Johnson's "Crossroads Blues," it was recognizable and true to the original but in an entirely new electrical context. It was played a good deal faster but within the classic twelve-bar blues form. In six all-instrumental choruses the three members of Cream demonstrated their skill at improvisation without leaving the form of the song. Unlike Johnson's original, Cream subdivided the beat into even parts and did not let the vocals depart from the beat established by the guitar. For many Americans who listened to it on the Cream album, this song was their first exposure to Johnson's music.

The English players had started by copying both the sound and the style of American bands; thus the Shadows did not alter or embellish the guitar sound of the Ventures' recordings, sticking to simple lead melodies picked out note by note with the rhythm accompaniment of strummed chords. But when it came to covering acoustic blues songs with electric guitars, several English players began to apply new technology to old musical ideas. Beginning with what they had learned from the bluesmen, Peter Townshend, Eric Clapton, Jeff Beck, and Jimmy Page looked to their amplifiers and effects devices to update the sound and make it meaningful in a new, and much louder, environment.

Beck, Clapton, and Page all came through the same band, the Yardbirds, which presented American blues pure and simple. The Yardbirds covered songs by John Lee Hooker, Howlin' Wolf, and Elmore James and did so with great respect and few changes. The sliding, wavering notes that the bluesmen had squeezed out of their battered old acoustics could now be created in the electric guitar's amplification and effects systems and reproduced at much greater volume. Modern blues guitarists also had the tremolo arm to get more sliding and bending of notes. Yet for all their blues credentials, the Yardbirds were still a pop band that issued pop singles, such as "Heart Full of Soul" (1965), which inevitably led to divisions within the group and the departure of blues purists. The tensions between pure blues and accessible pop might have been too much for Eric Clapton, but Jeff Beck and Jimmy Page were able to reconcile the two while never leaving the foundation of blues or the format of the three-minute song, such as "I'm a Man." This distortion-drenched recording was based on a blues standard but included some inspired guitar technique from Jeff Beck, such as controlled feedback, scratch-picking, and percussive drumming on the strings. "Shapes of

Things" (1966) showed how much he had explored the potential of feedback and distortion within the pop genre.

These English guitarists started by interpreting American roots music with American technology but were soon in a position to innovate with homegrown tools. Although there were no English guitar companies to challenge the leadership of the American manufacturers, there were several individuals who produced solidbody models, such as Charlie Watkins, Jim Burns, and Dick Sadler. Sadler's guitars were marketed by the United Kingdom's leading instrument wholesaler as the Dallas Tuxedo. Burns was a prolific guitar designer who earned comparisons with Leo Fender.

English inventors were much more active in amplifier and effects technology, and companies like Marshall and Vox could equal anything coming from the United States. Roger Mayer worked for the Admiralty in sound research but found enough spare time to build celebrated fuzz boxes for Jimmy Page and Jeff Beck. The Vox company emerged as a leader in effects technology, developing the wah-wah pedal. This was a device that simulated the open and close mute effect on a trumpet and made the sound that some compared to a baby's cry: "wah wah wah." Vox's engineers based the effect on a variable version of the midrange boost circuit used in their amplifiers. Significantly, this device was jointly developed between Vox and its American manufacturing associate Thomas Organ, which produced its own Cry Baby wah-wah.

The wah-wah sound effect was as critical to 1960s psychedelic music as distortion was to blues rock. Its popularity encouraged more entrepreneurs to build effects boxes and more musicians to buy them. The ability to create new electronic sounds from effects boxes was an important skill that had to be mastered by the guitar hero. The Yardbirds considered Jeff Beck to be their most creative guitarist, because he came with the technology in hand: "all these footpedals and fuzztones and feedback, something that we hadn't had before that."[7] Beck's Vox Tone Bender is conspicuous in his Yardbird recordings.

The one thing that distinguished the emergent British guitar heroes from their peers was that they played louder than anybody else. This was what first attracted attention to Eric Clapton and Jimmy Page. After one of his tours of the United Kingdom, Muddy Waters remembered that these guitarists "played louder than we had ever played."[8] In the 1960s the volume of the electric guitar rose exponentially as the makers of amplifiers listened to their most important customers. Hank Marvin and the Shadows liked their Vox AC15s but found that they did not have enough power to fill the bigger venues. After they completed a tour in the United States, they persuaded the Vox company to double the output—four power tubes instead of two—and the Vox AC30 was born. Even doubling the output was not considered adequate, and musicians continually complained that the AC30 was not loud enough.

The old format of having the electronics and one or two loudspeakers in the same box was soon deserted in favor of a separate electronic amplification unit (known as a head) running groups of loudspeakers. Once the output of the power amplifier exceeded fifty watts, designers faced the problem of blowing out speakers and employed several speakers to spread the load. Peter Townshend of the Who badgered Marshall to introduce a one-hundred-watt amplifier, the model 1959 Super Lead with four EL34 power tubes, that drove eight speakers in a massive cabinet. This proved too much for the roadies to carry, so Marshall divided the eight speakers into two cabinets—thus the famous Marshall "stack" was invented. The insatiable demand for more volume led guitarists to use multiple amplifiers and cabinets. Soon these stacks were taller than the musicians, towering above them in a wall of dull black boxes with a few tiny red lights glowing from the heads. This was a great contrast to the first amplifiers used with electric guitars, which sat sedately next to the instruments.

Throughout the 1960s and 1970s the trend on both sides of the Atlantic was toward even more powerful amplifiers; there were models on the market with names like Thunderstar, Dominator, and Exterminator and power peaks of 250 watts and beyond. AVT produced the ultimate bass amplifier with a power rating of 300 watts. The ideal amp was now one that "roars right thru anything in its path."[9]

The increased power of amplification had an effect on the composition of the typical rock band, which had been growing larger in the 1960s with the addition of rhythm guitarists and keyboard players. An alternative to the usual four- or five-man lineups was the "power trio" of electric guitar, bass, and drums, in which one very loud guitar did the job of both lead and rhythm. (In the studio they used double tracking or session men to add to their sound.) The power trio made up with volume and effects for what it lacked in manpower. This format focused more attention on the lead guitarist but also made the bass guitarist a star—Noel Redding in the Jimi Hendrix Experience and John Entwistle in the Who contributed a vital part to their bands' signature sound. The power trios could fill a large hall with loud music and entertain thousands of people. Cream's final concert at the Royal Albert Hall in London is a case in point. This magnificent auditorium had been built with large symphony orchestras and massed choirs in mind, but its cavernous space was easily filled with ear-deafening sound by three men playing in front of a huge stack of Marshall amplifiers.

Volume was changing the experience of listening to pop music. For the blues and folk purists, electronic amplification was an anathema. After Bob Dylan moved from acoustic to electric guitar, he faced considerable opposition during a famous 1966 concert tour (which followed his controversial Newport Festival appearance). Many of his fans objected to the loudness of the performance and claimed that it increased the separation of player from audience. When they heckled him during his performances, Dylan's answer was to play his Stratocaster louder. Muddy Waters was also harangued for playing the blues with an electric guitar. Ironically, at the same time that Dylan was touring in the United Kingdom, several African American bluesmen were also playing there with acoustic guitars, which their audiences interpreted as symbolic of authenticity. While Big Bill Broonzy and John Lee Hooker might have been willing to leave their electric guitars at home, Bob Dylan used his to demonstrate his independence from fans' expectations.

Loudness was now the goal: the manufacturers knew it as they scrambled to increase the output of amplifiers, and the audience heard it as they cowed in admiration before the noise coming from the stage. What were the roots of this obsession with volume? Anthropologists would point to the role that loud sounds make in the dominance-competition behavior of primates—the group to which monkeys, apes, and humans all belong. Primatologist Jane Goodall reported the story of the male chimpanzee Mike, who rose to become the dominant alpha male of his group in East Africa by means of an innovative noise-making display employing ten-gallon drums from the scientists' camp. Mike also inspired one of the younger chimps to imitate his behavior—something not unheard of in clubs and rock concerts.

A connection between maleness, dominance, and noise might be some 5 million years old, dating from the time of the first protohuman australopithecus, but the association of noise and assertiveness with masculinity nevertheless relies on a specific cultural context to derive meaning. Critics of rock 'n' roll have often seen it as sexist, and it cannot be denied that it occurred within a system of patriarchal power relations. Western notions of masculinity in the twentieth century equated power with sexual dominance and included loud display within the male rhetoric of supremacy. The value system of rock 'n' roll and the technological capabilities of the electric guitar fitted in quite nicely with these cultural constructions and reinforced the association of high volume with assertive male behavior.

the guitar hero and masculinity

It was no coincidence that in the 1960s all the guitar heroes were men and that the life of a rock 'n' roll star was an advertisement for testosterone-driven behavior. The bluesman figure was a modernization of the romantic ideal of the hero in European culture, in which the hero was always male and the females were either admiring facilitators (muses) or obstacles (see plate 8). The guitar-playing rock star might have been an updating of the romanticism of early modern Europe, a continuation of the Romantic ideal in the new world, but the instrument itself had suffered a role reversal.

For centuries in European culture women were most associated with guitar playing. From the sixteenth century onward the guitar was normally depicted as being played by a women and having a sexual connotation of femininity. In the 1630s the French writer Pierre Trichet concluded that the guitar "had a certain something which is feminine and pleasing to women . . . making them inclined to voluptuousness." A seventeenth-century guitarist wrote that "a guitar is a woman [who] will not wither, no matter how much you touch her with your hands."[10] In the nineteenth century the guitar remained intensely popular with women and was continually represented as an instrument that a woman played in the tranquility of her own home. A German observer wrote in 1804 that the guitar could be found in the home of every woman: "modern, attractive, affectionate, flirtatious, playful, pretty, exuberant, mischievous or even innocent, demure, respectable women."[11] Innocent young women, fashionable ladies, and temptresses were all depicted playing the guitar in what remained a common theme in the visual arts.

But in the twentieth century the guitar's cultural symbolism changed radically as the forces of technology and modernism transformed ways of life (see plate 9). Electrifying the instrument was the decisive step in making the guitar a symbol for masculinity. Electricity was one of the important symbols of modernity in the United States and western Europe, and it was entirely in male hands. Technology was considered a male preserve, for there were very few famous female inventors, and even technological enthusiasm was seen as an activity of boys and young men. In the hands of inventors and engineers, electricity had become an obedient tool of mankind, but it also had a sexual meaning. Early in the twentieth century attraction between men and women was described in terms of electrical energy; flirting was "sparking." In 1901 the Sears Roebuck retail firm promised customers that the "Heidelberg Electric Belt" would cure "disorders and weaknesses peculiar to men," restoring "superb manliness and youthful vigor."[12]

The Heidelberg Electric Belt might not have done the trick, but fifty years later the electric guitar on stage certainly did. In the world of rock 'n' roll the guitar was an inescapable symbol of masculinity, and the dynamics of the performance were filled with sexually significant actions and meanings. The electric guitar became part of the rite of passage of the adolescent male, either literally or figuratively (the air guitar), and handling the guitar became an expression of masculine aggression and power. In several genres of rock, especially heavy metal and hard rock, the way of holding and moving the guitar was closely connected to its phallic symbolism. The erect guitar became an essential part of the formalized ritual of the rock concert. In the hands of rock musicians the guitar was a symbol of controlled power as its volume reached unworldly levels. When the guitar virtuosos played with feedback, they demonstrated that even the most errant and unpredictable forces of nature could be brought under control. At home or on stage, strumming a real or imagined guitar gave its youthful player a feeling of enhanced energy and potency. In this way the meaning of the sound meshed with contemporary notions of masculinity.

The electric guitar was seen and understood as a boy's toy, a part of the paraphernalia of growing up male, like a Red Ryder airgun or a football. Behind this commonly held perception was the reality of the economic and technological basis of popular entertainment. The business organizations that made electric guitars and marketed their music were run by men. There had been some outstanding women blues guitarists in the 1930s and 1940s, and there were working female guitarists in the 1950s and 1960s; but there were no women running record companies, operating recording studios, or managing the factories that mass produced electric guitars. The electric guitar operated in a male domain.

The guitar hero strengthened the ties between masculinity and the instrument, and the arrival of the ultimate sixties guitar hero marked a climax of sexuality and virtuoso playing. All the elements of the bluesman myth came together in Jimi Hendrix—the virtuosity, the tragedy, the sexual power, and the self-destruction. On one hand he was the most technologically adroit of all the guitar heroes, but on the other his music and life had an air of mystery and magic. In Hendrix the European and West African antecedents of the hero came together. He mixed his blues roots with the advanced techniques of distortion and feedback and had the confidence to embark on long jazzlike improvisations. Guitarists like Hendrix were stringing multiple effects boxes together: fuzz tone, treble booster, echo relay, and wah-wah. He played his amplifier and effects boxes as much as he played his guitar. When he first started playing in England in 1966, almost every guitar player or studio technician who heard him said they had

Jimi Hendrix in full psychedelic regalia playing his Fender Strat. Being left-handed, he played his guitar upside down. From the collection of Rock and Roll Hall of Fame and Museum.

never heard sounds like that before. His virtuosity was unparalleled and his sound began to characterize the music of the psychedelic era. When Jimi Hendrix told an audience, "You'll never have to hear surf music again!" he was bringing an end to a period when amateur guitarists felt that they too could be heroes: no one could imitate the playing of Hendrix, and the line had been drawn between amateur and virtuoso.

Whereas Beck or Clapton had utilized feedback for the guitar solo or for a flourish, Hendrix's music could be all feedback. He controlled and manipulated it, exploring all its tones and dissonances to find music in it. The distortion was so severe in his first record, "Are You Experienced?" that the record presser returned the masters as defective. Many people who listened to it thought it was dubbed-in electronic music, not the sound of a Fender Stratocaster. His wailing, distorted version of the "Star-Spangled Banner" on the final day of the Woodstock festival is a masterpiece of controlled distortion and subversive music making. Hendrix was the master of sexually inspired guitar spectacle. While heroes like Clapton and Townshend had tapped into the persona of the bluesman, Hendrix was the real thing—a black man playing the blues. The myth of African American sexual prowess was there to be exploited, and Hendrix's physical interaction with his guitar—rubbing it against his crotch, bumping and grinding on it—emphasized the sexual power of the guitar player. The guitar had become, in the words of Steven Waksman, "the instrument and the symbol for a highly gendered and racialized form of virtuosity."[13]

The significance of the electric guitar as a symbol grew with the size of the spectacle it powered. The rock 'n' roll concert became a major event in the 1960s. Bands were playing bigger venues such as sports stadiums, and the old tradition of having numerous acts singing a few songs each was dropped in favor of a warm-up act followed by the headliner, who played for hours. The really large concerts, such as the Woodstock festival or the Monterey Pop festival, have become historic events, often interpreted as important turning points in cultural history. It was at Monterey that Jimi Hendrix supplied the ultimate ending to the guitar-based spectacle; after playing his version of "Wild Thing," he poured lighter fluid on his guitar and burned it in front of an amazed crowd.

Hendrix might have been driven to this costly piece of stagecraft because he knew he had to follow the Who on stage, a band that had already made the destruction of its equipment an essential and expected part of its show. Rock critics and fans of the band have come up with several different explanations for this practice, but in Townshend's own words, "Every night we wanted to go a bit further, but we could not because we did not have any more volume, musical ideas or musical dexterity."[14] Thus when the guitarist reached the limit of high volume and virtuosity, the only thing left to do was to destroy the guitar and amplifier in a symbolic act of frustration. The world of chaos and anxiety explored by the guitar heroes then came to a fiery, apocalyptic end. The tragic ending was an important part of the guitar-hero myth, and the life of the modern bluesman, like that of his predecessors, was not without risks: Hendrix died from an overdose; Clapton and Townshend both suffered hearing loss from playing so loudly. Eric Clapton remembered: "I was playing at full volume in a kind of weird traumatic state."[15] The deaths and the breakdowns only added to the mystique of the guitar hero and created more martyrs for the music.

Sources

Foster, Mo. *17 watts? The Birth of British Rock Guitar.* Bodmin, U.K.: MPG Books, 2000.
Guitar Player and *GuitarOne* magazines.

Notes

1. Paul Oliver, *Blues Fell This Morning: Meaning in the Blues* (London: Cambridge University Press, 1990), p. 107. "Sensational pickup" quote is by John Whetton and appears in Foster, *17 Watts,* p. 130. Kenny Wayne Shepherd quote in *GuitarOne,* December 1999, p. 68.

2. Hank Marvin's Stratocaster acquisition is from Foster, *17 watts,* p. 163: Townshend quote is from his liner notes (written in 1983) to *Twang: A Tribute to Hank Marvin and the Shadows,* Pangea Records, 1996.

3. The musicians referred to were George Harrison and John Lennon. Tony Bacon and Paul Day, *The Rickenbacker Book* (San Francisco: Miller Freeman, 1994), p. 29; the dream quote is from Brian May, in Foster, *17 Watts,* p. 108.

4. Charlie Gillett, *The Sound of the City: The Rise of Rock and Roll* (New York: Pantheon, 1983), p. 384.

5. Quoted by Charles Shaar Murray, *Boogie Man: The Adventures of John Lee Hooker in the American Twentieth Century* (New York: St Martins, 2000), p. 368. See also George Lipsitz, *Dangerous Crossroads: Popular Music, Postmodernism, and the Poetics of Place* (London: Verso, 1994), chap. 3.

6. Buddy Guy, *Guitar Player,* February 1969, p. 21.

7. Liner notes, *Beckology,* Sony Music, 1991, p. 10, quoting Jim McCarty.

8. *GuitarOne,* October 2001, p. 88.

9. *Guitar Player,* June 1969.

10. Quoted in Frederic V. Grunfeld, *The Art and Times of the Guitar: An Illustrated History* (New York: Da Capo, 1974), pp. 30, 98, 106.

11. Ibid., pp. 115, 169.

12. David Nye, *Electrifying America* (Cambridge: Harvard University Press, 1990), pp. 154, 156.

13. Steve Waksman, *Instruments of Desire* (Cambridge: Harvard University Press, 2000), p. 201.

14. Peter Townshend, *Guitar Player,* October 1967, p. 8.

15. Eric Clapton, *Guitar Player,* June 1970, p. 40.

Further Reading

Gordon, Robert. *Can't Be Satisfied: The Life and Times of Muddy Waters.* Boston: Little, Brown, 2002.

Levine, Lawrence. *Black Culture and Black Consciousness.* New York: Oxford, 1977.

8 heavy metal

from guitar heroes to guitar gods

André Millard,
after the collaborative paper
with Rebecca McSwain

The guitar hero of the 1960s was a reflection of the dominance of guitar-based rock in the popular music of the times. This hegemony was challenged in the 1970s, with the increased availability of electronic synthesizers and other electronic sound generators, such as drum machines. The inventor Robert Moog began marketing his synthesizers in earnest, and gradually this increased the influence of keyboards in rock bands. By the end of the decade this new electronic sound had grown so popular that die-hard rockers were labeling their recordings "synthesizer free." As new electronic sounds permeated popular music, new alternatives emerged to challenge the format of guitars and drums. Disco did away with the musicians altogether with an entertainment of recorded sound assembled by DJs. Instead of the vocalist or the guitar hero, the dancers were the stars. The 1970s might be remembered as the decade of disco and the electronic synthesizer, yet the electric guitar did more than survive; it became enshrined as the central sound and the unmistakable icon of popular music, reaching beyond its bastion in youth culture and pop songs to infiltrate commercials and film music.

The belief that the future of rock 'n' roll was still going to be a man with a guitar was reaffirmed when Bruce Springsteen appeared on the covers of both *Time* and *Newsweek* in October 1975. Springsteen was a rock traditionalist, and he was shown with an electric guitar in his hand. The steady development of technique and technology brought new levels of flexibility, power, and speed to guitar music, and subsequently, in the words of Robert Walser, "the electric guitar acquired the capabilities of the premier virtuosic instruments of the seventeenth and eighteenth centuries."[1]

As the audience for pop music became larger and more diversified in the 1970s and as bands directed their sound to narrower segments of the music-listening population, rock 'n' roll 1950s style evolved into several different strains. Record companies and radio stations broke down American youth culture into different audiences, each with its own genre of music. *Rock 'n' roll* as an identifying term was divided into separate subcategories, normally expressed as *rock* combined with a descriptive adjective: folk rock, psychedelic rock, and so on. One of these categories, heavy metal, took guitar admiration and the symbolic power of this instrument to new levels.

The term *heavy metal* is mentioned in a book by the avant-garde writer William Burroughs, and it first occurred in a song's lyrics in Steppenwolf's "Born to Be Wild" from the soundtrack of the quintessential sixties counterculture movie *Easy Rider.* It was being used at the start of the 1970s to describe the loud blues-based music of bands like Led Zeppelin and Black Sabbath. Whatever its provenance, the term is significant; it is based on the sound, not geographical orientation (southern rock) or inspiration (folk rock) or aspirations (art rock). With its close counterpart hard rock, it implies louder and more aggressive sounds. *Heavy* means loud, with the emphasis on the lower registers of the bass, and *metal* indicates the harder, sharper sounds of the solidbody electric guitar. As Harris Berger found when he interviewed metalheads in northeastern Ohio in the 1990s, the fans considered that rock had long ago lost the mantle of *heaviness*—a term that encompasses the sound, the seriousness, the meaning, and the worth of popular music.

The heavy metal sound had its roots in 1960s surf music, such as Link Wray's "Rumble," and in what would now be called the power pop of the Kinks and the Who. Recordings such as "You Really Got Me" and "I Can't Explain" infused powerful, dense guitar sounds with excitement and drama. Heavy metal originated with the guitar heroes of the 1960s. Townshend, Hendrix, Clapton, Beck, and Page perfected the power chord (the stripped-down basic building block of a root, its 5th, and an octave), used feedback and distortion as a form of musical expression, and played really loudly. Their music was based in the blues, but they played a variety of music, both light and heavy.

The band that made the transition between the 1960s power pop sound and the new sound of heavy metal was the hard rock band Led Zeppelin. It had a virtuoso guitarist (Jimmy Page), a strong foundation in the blues, and, most important, a commitment to volume that was unprecedented. For these reasons Led Zeppelin is usually the starting point of any history of heavy metal. The name itself suggests opposing forces—light and heavy, acoustic and electric. Led Zeppelin's famous "Stairway to Heaven" begins with an acoustic guitar and then works itself up into a frenzy of amplified guitar and distortion.

Fans of heavy metal still debate the merits of Led Zeppelin as pioneers of the sound and usually distance themselves from the more commercial and thus compromised hard rock. There is far less disagreement on the position of guitarist Tony Iommi of Black Sabbath as a founding father. An accident resulted in the loss of the tips of his fingers, and subsequently he tuned down his guitar a half step to reduce the tension of the strings and make them easier to play. This gave his guitar a lower, heavier, thicker tone. Iommi also simplified the chords to two- and three-finger power chords. This does not mean he made the mechanics of playing any easier—far from it—nor did he desert complex melodies, but he did strip down the sound and make the repetitive guitar riffs that propel many Black Sabbath songs more sparse and powerful.

If we position heavy metal as an offshoot of blues-based hard rock, it is not surprising that the first bands to be labeled heavy metal came from England. This important new guitar sound had its roots in the transatlantic diffusion of culture and technology. Black Sabbath, often referred to as the originator of the music, was formed in Birmingham, England, in 1969. Deep Purple and Uriah Heep also came into being there about this time. Each of them had a lead guitar virtuoso: Tony Iommi, Ritchie Blackmore, and Mick Box, respectively. In the late 1970s another wave of British bands, such as Iron Maiden and Judas Priest, found a large audience in the United States and helped establish metal as a solid commercial genre. Heavy metal bands were also formed outside the traditional nexus of America and the United Kingdom: AC/DC in Australia and the Scorpions in Germany.

Deena Weinstein describes the metal movement as evolving into two strands in the 1980s: "lite" metal on one hand, and speed or thrash on the other. The former includes successful recording artists with elaborate shows, such as Van Halen, Def Leppard and Quiet Riot, while the latter would lead to bands like Metallica and Megadeth—equally successful in selling records but with a much darker outlook in the lyrics and more conservative performances. The lite metal acts dominated the concert scene and sales of recordings in the 1980s. Def Leppard's "Pyromania" sold 10 million copies, second only to Michael Jackson in record sales in that decade. The wide appeal of bands like Van Halen and Guns

N' Roses completed the guitar's hegemony of the 1980s when five heavy metal acts sold over 30 million records: proof enough that this sound had entered the mainstream of popular music. The fans of heavy metal and hard rock explained this popularity as a reaffirmation of the integrity and excitement of an earlier era: they were enjoying the "true" rock 'n' roll of the 1980s.

Often reviled by the critics and criticized by anxious parents, heavy metal became a significant presence in the popular music of the late twentieth century. The commercial acceptance of lite metal (as expressed in record sales and radio and television exposure) marked the triumph of the electric guitar in the American century. The popularity of heavy metal as a whole made sure that the electric guitar survived the competition of other instruments and changing musical fashions in the 1980s and 1990s to remain the signature sound of youth music. If one took Woodstock 99 as a summation of the popular music on the eve of the new millennium, one could only conclude that the heavy metal guitar and its sound were as pervasive as ever: the booming volume, massive distortion, and crunching riffs all pointed back to a previous era when the guitar player was king.

the metal sound

The heavy metal sound is fundamentally the sound of the electric guitar amplified and distorted to the limit. The most important criterion is that it has to be loud, very loud—louder than any other music. Although other types of rock musicians wanted to play loud, heavy metal players took the power of electric amplification to its ultimate point. In the early days of the solidbody electric guitar, the quest for more volume was practical and commercial; the guitar had to be heard in a large band. After Jimi Hendrix pushed through the sound barrier by controlling the feedback he created, loudness became a virtue in its own right.

Led Zeppelin raised the barrier on what was considered acceptable as a volume level in its concerts. When Black Sabbath was touring the United States in the first years of the 1970s, its sound was promoted as "louder than Led Zeppelin." It was followed by Deep Purple, which made the *Guinness Book of Records* with its output of 117 db—only 3 db short of the pain threshold and equal to the sound of a jet aircraft at one thousand feet! With exhortations in their songs to "crank it up" and "blow out the speakers," heavy metal bands have pushed the envelope of how much volume they can generate and how much their audience can stand.

This was not just an effect; it was a vital part of the experience of the music. As one fan put it, "the whole point of heavy metal music is to get out of your mind. The music is always so loud I can never hear the words, but it is just basic noise to blast your brains out." Intense volume brings in more than just an intellectual response to the music, it makes it feral and emotional. Described as "a series of sonic body blows, true violence and trauma," it is a sound so loud that it "purifies" the mind, overcoming reality with noise.[2]

Heavy metal would have been technically impossible in the 1950s, because guitar amplifiers were not powerful enough to produce the volume and distorted timbres that are its trademarks. Guitarists in the 1970s had more power at their fingertips than anybody before them and the excuse to play even louder. Manowar's song "All Men Play on Ten" refers to the setting on the amplifier control panel. In the film parody of heavy metal music, *This Is Spinal Tap,* a guitarist in the mythical English band Spinal Tap goes one step further and has the volume settings on the amps set from 1 to 11, giving an illusory extra 10 percent of power. The Marshall Company used the actors from the film to promote their products, and the band actually toured the United States. Heavy metal myth thus achieved some measure of reality.

Extremely loud music forces the guitar player to simplify. As Eddie Van Halen said, "you start dickin' with chords like 7ths and 9ths through a blazing Marshall, and it will sound like crap."[3] Harmonies extended beyond the triad might sound great on an acoustic, but crank up the power and the harmonics and interplay of sound get lost in the noise. Simplicity can be found in the basic power chords that require only two or three fretted strings. It can be heard throughout the AC/DC catalog, for example the three open-position power chords (E5, D5, A5) in "Back in Black," and in Ritchie Blackmore's sludgelike tone that characterizes Deep Purple's "Smoke on the Water."

Pitch is critical to the heavy metal sound; it puts the heavy into heavy metal. *Heavy* refers to the bottom end of the low frequencies, the throbbing bass notes that can be felt in the chest, something that can be physically experienced by the body rather than just heard. These low frequencies need more amplification if they are to be heard as the overall volume of the sound is raised, and this puts further demands on the amplifier. While the emphasis is on the lower register, especially for the underpinning riff of many songs, heavy metal music also reaches for the highest of highs, contrasting the bass of the rhythm with the soaring solo notes of the lead guitar.

Distortion is another fundamental of the heavy metal sound. Any increase in the amplification of sound brings the potential of increased distortion—the higher the volume, the more demands on the amplifier and the speakers. Distortion in heavy metal is seen as highly desirable, not just for the sound of the guitars, but in the wailing and screaming of the vocalists, which suggest emotional turmoil. Distortion also extends the spectral range of the guitar. It adds exciting new timbres to the sound, and it also gives that unworldly, alien effect that goes into the making of a guitar god: "the more distortion you get the more satisfying it is. There's something slightly superhuman . . . about the sustain, the nearly endless notes."[4]

The heavy metal virtuosos demonstrated their calling in the guitar solo, which became an obligatory part of the music and a fixture in nearly every song. In the late 1960s guitar solos got longer and more involved. This was because the artists were more self-confident, the audiences were more indulgent, and conversion of the recording format from three-minute single to forty-minute long player gave musicians a lot more time to develop their ideas. Some of the best-loved hard rock songs, such as "Freebird" and "Stairway to Heaven," were built around the solo, and in some cases the solo took over the song. The guitar solos that occur within a deafening wall of sound cannot be the same as the smoothly assembled solos of jazz, for example, yet they are expected to be just as precisely articulated. Heavy metal guitarists like Randy Rhoads and Yngwie Malmsteen managed to construct complex solo lines within the context of blazing distortion and high volume.

At the end of the 1970s a new wave of heavy metal guitarists emerged, including Rhoads, Van Halen, and Malmsteen, who dazzled their audiences with amazing technique and flashy showmanship. They took earsplitting volume and dizzying speed to the limits of human endurance but still produced some of the most formally complex music in the heavy metal canon. They developed new techniques to accomplish their ambitious musical goals, such as the technique of tapping the frets with a finger (on the right strumming hand) while playing. Tapping just behind the fret enables the player to produce long, smooth legato lines, which were impossible with conventional left-hand fingering.

While heavy metal as a whole got a reputation as a loud and simplistic music, played and enjoyed by the morons satirized in *Spinal Tap,* the guitar virtuosos from Ritchie Blackmore to Eddie Van Halen were constantly improving their technique. In their hands the electric guitar was the means to create a formally complex music, as important as the piano was to the nineteenth-century composers. That might explain why several heavy metal guitarists have a genuine interest in classical music. Ritchie Blackmore, one of the masters of heavy metal guitar, was first inspired by the sound of Duane Eddy and then mightily impressed with the Who and songs like "My Generation" as a teenager, but he also maintained a strong interest in Bach and Beethoven: "I loved the drama in music and the classical side of rock."[5]

Renewed interest in the electric guitar as a virtuoso instrument brought instrumental songs back to the *Billboard* charts for the first time since the garage bands and surf music of the 1960s. Complex guitar instrumentals by Jeff Beck and Joe Satriani achieved widespread popularity in the 1980s, and their albums *There and Back* and *Surfing with the Alien* made the Top 40. But unlike the instrumentals of the 1960s, the playing on these songs could not be easily duplicated at home, and that was the point—it was only for the master musician.

Heavy metal put a premium on virtuosity and made sure that the electric guitar remained the vehicle for the expression of heroic individualism. The guitar hero of the 1960s mutated into something even more powerful, more romantic, and more adulated. Heavy metal boasted the most committed fans in the whole world of rock—the "headbangers" or "metalheads" who turned out in thousands for the show and who made the music a vital part of their identity. The heavy metal concert allowed fans to demonstrate their allegiance to the music and worship the stars on the stage. In this way the guitar heroes of the 1960s were joined in the pantheon by a new breed, similarly worshiped but using the electric guitar in new ways, defining a different universe. The new hero was no less adept a player, but now he was a costumed, often cynical, sometimes angry figure.

While the sixties heroes were certainly virtuosos, they appeared primarily as working men, musicians who made a living with their instruments: "I still got my electric guitar," Jimi Hendrix says in his songs when everything else goes wrong. The sixties heroes were earthbound (that is, until their untimely and tragic deaths), and they did not appear on stage through billowing smoke as a visitation of a heavenly object. Pop music was achieving a new sense of spectacle in the 1970s, and the rock guitarist was becoming a distant, supernatural star rather than an (extremely talented) guy next door. Eric Clapton's fans might have thought of him as a god, but the unassuming musician certainly did not act or perform on stage like one. On the cover of the seminal John Mayall and the Bluesbreakers album he is shown reading a children's comic. Heavy metal guitar virtuosos were equally admired, but they tended to act the part, especially when they achieved a level of fame and affluence that was excessive even by rock star standards. As Ozzy Osbourne said of Black Sabbath when that group first tasted success: "We started to think we were gods."[6]

In the rarified concerts of the 1980s the guitar god was the central object of veneration. Both life-affirming and larger than life, the prancing guitar player paraded the sexual energies of his audience while demonstrating unprecedented control and power over the instrument in his hands. He had the ability to transcend the moment and take his audience with him, bringing them to a mysterious, creative world beyond time and place where perfection is possible, if only for a brief moment. But the place we were taken to by the electric guitar virtuosos was a new one, in which boundaries of sexuality, social structure, and power were deliberately confronted, mocked, and outraged. In the words of Davie Bowie, describing his mythical guitar god Ziggy Stardust, he was "a leather messiah."

heavy metal guitars

Heavy metal drew on both the technological development of guitars and amplifiers and the sociocultural symbolism of the machine to its audience. While the sound depended on the loudness and distortion produced by amplifiers, the show and the creation of guitar gods owed a lot to the power of the electric guitar as a symbol. There were new fashions in the look of the guitar. The rise of hard rock and heavy metal also revived the fortunes of several guitars that had fallen into disfavor.

The historic Les Paul models were discontinued by Gibson in 1961. The replacement was an entirely new design, which had twin hornlike cutaways. It did not get the endorsement from Les Paul and was eventually named the SG (for standard guitar). Thinner and lighter than the old Les Pauls, the SG had a flatter-radius fingerboard, which meant it was easier to play fast. Eric Clapton played one when he was with Cream, and the SG found favor with some of the revered pioneers of heavy metal guitar, such as Tony Iommi and Ritchie Blackmore. In the 1970s some of the new generation of hard rock and heavy metal guitarists, notably Angus Young of AC/DC, also adopted it.

The old Les Paul Standards and Goldtops from the 1950s were still prized by young guitarists like Eric Clapton (when he played in John Mayall's Bluesbreakers) or the American Mike Bloomfield with the Paul Butterfield Blues Band. The Goldtops with their P-90 single-coil pickups were heavily favored by blues guitarists, including Muddy Waters, John Lee Hooker, Freddy King, and Guitar Slim. The Standards, on the other hand, were not as popular and thus their secondhand prices were lower. But they had humbucker pickups and this made them ideal for the louder volume of blues rock. Combined with a suitable amp, such as the Marshall JTM-45, the Standards had the power and tone to handle high-powered, blues-based rock. A thick and bottom-heavy sound, combined with high levels of sustain, made these guitars the ideal instruments for hard rock and heavy metal. In the words of one commentator, heavy metal was "the house of pain that Gibson and Les Paul have wrought."[7]

The chunkier, growling tone of the humbucker pickups were easily distinguished from the clearer and more penetrating sound of the single-coil pickups of Fender guitars—the sounds that had become associated with surf and guitar instrumental recordings. The Gibson tone—fat and distorted—became fashionable in the 1970s, and Les Pauls became the guitar of choice for many rock guitarists, from heavy metal guitar gods to southern rockers. As the Atlanta Rhythm Section sang, "lay down the back beat / crank up the trusty Gibson." From Jimmy Page and Duane Allman to Ace Frehley of Kiss, everybody was strapping on a Les Paul. By the 1980s pop music was awash in humbucker guitar tone (usually in the bridge position) driven through an overloaded tube (or tube-sounding) amplifier. The old reliable Stratocaster and its cleaner tone fell into disfavor.

The problem was how to acquire a Les Paul now that it was out of production. There were plenty of copies on the market, as many as thirty by the late 1970s, and the Gibson company was also marketing reissues (which were not exact replicas of the originals), but musicians were not satisfied with the quality of American guitars produced in the 1970s. The players created the first market for what are now called "antique" (instead of "used") guitars by buying secondhand instruments rather than new models. Their demand for the classic 1950s and 1960s models drove prices up during the decade; a pre-CBS 1957 Strat that could have been picked up for a few hundred dollars at the beginning of the 1970s ended up around $1,000 ten years later. Les Pauls were in favor and thus more expensive—prices rose from around $500 to close to $2,000.

Heavy metal players were an important part of the market for antique guitars. Most of the metal virtuosos had collections of guitars—not one or two, but rooms full of valuable instruments. This, too, was satirized in *Spinal Tap*. A collection of classic instruments was now considered one of the perks of guitar stardom, along with the private jet, limousines, and a special helper to take care of the tools of the trade—the guitar tech.

While the players were looking back over their shoulders at the time-honored classic guitars of 1950s rock 'n' roll, the manufacturers were usually looking forward into the future and the promise of technological progress. Doubtless encouraged by the players' demand for more power, they were focused on a systematic improvement of the technology of electronic amplification. They were thinking in terms of output and efficiency and attempting to innovate their way out of the stagnant market for new guitars in the 1970s.

The first stop in the search for more output had to be the pickups. Hard rock and heavy metal players wanted high-output, "hotter" pickups that generated more volume. Designers examined more powerful magnetic materials and experimented with elaborate winding techniques to give more output. They developed active (instead of passive) circuits in the pickups that gave the player the means of processing the signal before it left the guitar. Instead of just running the signal to the amplifier, engineers devised ways of processing it onboard the guitar. In this way the preamplifier section of the amplifier was built into the guitar and controlled by the player. This technological innovation was incorporated into the Gibson Les Paul Recording guitar, which had a control panel at the bottom of the body that worked (and looked like) a recording studio console. It gave the guitar player the same control over the output as the recording engineer had in the studio. Billed as "The First Computer like Guitar" it was a commercial failure.[8] The Ovation Breadwinner (1972) had a preamplifier built into it, and it was the first guitar with active electronics to be mass produced in the United States. The Gibson Artist RD (1978) had active circuits designed by the Moog synthesizer company.

Active pickups were a major innovation that rested on the advances in solid state integrated circuits. Their small size, low heat output, and superior performance made it possible to duplicate the bulky boards in the amplifiers with much smaller units that could fit into a guitar's body or into an effects box. Solid state gadgets built into guitars were a predictable development in electric guitar technology, as were new materials for the solid body. Ovation, the company that produced active circuitry guitars, was founded to explore the use of composite materials in guitars. Yet the natural development of electronics and solidbodies went far beyond what the average guitarist thought appropriate. The majority of them shunned guitars with the effects built in for stand-alone effects boxes or pedals that could be purchased one at a time. Consequently, a new generation of "smart guitars" did not catch on.

Ironically, at the time when heavy metal players were enhancing the status of the instrument, the manufacturers were facing one of the most disappointing decades in their history. After the amazing growth of sales in the 1960s, the American guitar companies found themselves in an arena where the two forces of technology and spectacle collided, and they failed to understand or reconcile the two. They might have understood the technology, but they often failed to anticipate how the players would receive and use it.

As the great guitar boom slowed down in the 1970s, American manufacturers had to face an increase in foreign competition, especially from Japan, that cruelly exposed their inadequacies. In an article in *Guitar Player* in 1979 entitled "The Future of the American Guitar," the guitar historian Tom Wheeler posed the relevant question: "Does it have one?" The leading American guitar manufacturers introduced many new models in the 1970s, such as the Fender Starcaster, with very little success. During this period the competition from Japan moved from low-price amateur models to high-quality professional equipment. Yamaha, an old established maker of acoustics, started to produce electrics in 1960s and by the mid 1970s was producing guitars—the SG 2000 series—that could equal the quality of the best American models.

The penetration of Japanese electric guitars into the American market started in the 1960s with cheap copies of Fender and Gibson guitars. Naturally the Stratocaster and the Les Paul provided the model for many designs, but the Japanese were also open to other inspirations, such as the Mosrite, which was well known in Japan because of the popularity of the Ventures. Japanese manufacturers often took the spirit rather than the detail of the original and incorporated ideas from many different designs into one guitar. Despite their reputation for slavish copies, the Japanese showed that they were open to innovation, especially in body shapes, new sparkle finishes, and unusual colors. As the history of the Fernandes company bears out, making replicas of vintage American guitars worked fine for a while, but if the company was to grow in the United States, "unique designs and concepts were required to give Fernandes its own identity."[9]

The rising value of the Japanese yen drove up manufacturing prices and forced Japanese companies to look for sales at the higher end of the market, where they had to compete on the basis of quality and innovation rather than with cheap copies. (It also made manufacturing outside Japan more attractive.) As they were not bound to traditions of making guitars, Japanese companies could experiment with new shapes for the body and headstock and with new materials, such as fiberglass and aluminum.

The leading exporters were Tokai, which produced a popular Mosrite look-alike and the metal-bodied Talbo model, and Fuji Gen Gakki, which sold its products as Goldtone guitars in the United States. The Hoshino company's Ibanez guitars appeared in the 1970s as competent copies of several American guitars, including the novel Ampeg Dan Armstrong clear acrylic guitar—a reworking in cast plastic of a classic 1960s instrument. The Ibanez Tube Screamer effects pedal, which emulated the sound of an overworked vacuum tube amplifier, was also a major success for the company. Ibanez produced guitars with unusual styling—elongated and twisted cutaways, spaceship shapes, and reversed sharp-pointed headstocks. Such was the growing influence of Japanese manufacturers that this latter design innovation was referred to as the Samurai (sword) headstock. It was widely copied in the 1980s. The Ibanez 1979 Iceman looked outrageous, with exaggerated, sharply defined cutaways, but it was a spectacular success with young players. Endorsed by Paul Stanley of Kiss, it has become a hard rock classic, and Ibanez manufactured reproductions of this model in the 1990s.

Ibanez was joined by several other manufacturers that produced unusual and unique-looking guitars. Formed in Japan in 1953, the Aria company began by making acoustics and branched out into electrics in the 1970s. The Aria Urchin (1977) looked nothing like an electric guitar, except that it had a neck with strings; the neck was attached to a shape that looked like something from outer space. Other Japanese guitars from Washburn (once an American concern) and ESP carried on this tradition of novel and arresting styling. Some of the strangest-looking instruments were marketed under the Hondo name, which was a collaboration of the International Music Corporation (based in Texas) and the Korean manufacturer Samick. The Hondo Longhorn bass guitar took the exaggerated horned body to the extreme.

The most influential guitar shape of heavy metal and hard rock guitars came from the past rather than from the imagined future. The Gibson Flying V, Moderne, and Explorer had been a brave journey into a new world of guitar shapes that had not translated into commercial success in the 1950s. Very few were made and even fewer were sold. In the 1960s Gibson again attempted a radical new look with the Firebird models, which were designed with the help of automotive stylist Ray Dietrich. These too were not smash hits on the guitar market. Yet interest in the unusual Flying V led Gibson to reintroduce a line of reproductions in the late 1960s. Flying Vs were attractive not only to hard rockers; veteran blues player Albert King used one to achieve his distinctive sound. There were so few Flying Vs manufactured that the prices for used models were some of the highest in the antique guitar market, usually well in excess of two thousand dollars. This encouraged small companies to produce copies of the classic V. When guitar designers in the 1970s were casting around for a new look, they were attracted to the unusual shape of the Flying Vs and Explorers. The obtuse lines of the Explorers, angular and dissonant, inspired a body shape that fascinated many metal players. With new finishes and colors the familiar zigzag body shape became standard equipment in hard rock and heavy metal bands. The Explorer and the Flying V shapes are most closely identified with guitars of this genre.

The popularity of loud guitar music encouraged new entrants into the American manufacturing industry. Dean (formed in 1979), Hamer (1975) and B.C. Rich (1979) began as repair shops or dealers in antique guitars. Hamer was founded by Paul Hamer and Jol Dantzig, who started a business repairing Gibson electrics. Their Northern Prairie Music store in Chicago was one of the first to deal in antique guitars and became a popular stop with many touring bands. They decided to move into the manufacture of a few classic Gibson models, but with the highest-quality materials and with authentic vintage hardware they picked up at the Gibson plant. Their 1974 Standard was based on the Explorer and went into limited production. Rick Nielsen of the hard rock band Cheap Trick helped Hamer achieve prominence when he adopted one of their first Standard guitars. A list of their earliest customers indicates the type of music being made with their products: Bad Company, Deep Purple, Def Leppard, and Kiss.

Dean Zelinsky was another young man running a repair shop and used-guitar business in the Chicago area. He too moved into making faithful copies of classic Gibsons. The demand for these reinterpretations of older designs made Zelinsky think that the established guitar manufacturers were out of touch with rock guitarists. "I saw a huge generation gap," he recounted, "and all the technological fads and new electronics were not in step with what these players wanted."[10] Dean Custom guitars soon moved into making limited production runs of original designs inspired by Flying Vs and Explorers, such as the ML, Z, and E'lite guitars. These models had an oversized V-shaped headstock, which became identified with the brand.

Kramer guitars also had an unusual split headstock. Gary Kramer was a partner of Travis Bean in the manufacture of aluminum-neck guitars. In 1975 the Kramer company was formed; it produced several versions of the V and Explorer shapes, sometimes inverting the latter to produce a novel design. The XL series of guitars had some very unusual shapes, such as a shark's-fin cutaway, and in 1986 the company produced a '57 Chevy tail fin model. Kramer was certainly in tune with the needs of hard rock and heavy metal players, producing guitars with high-output pickups and locking tremolo units. The classic V and Explorer guitars have survived well into the 2000s with Washburn, Dean, Jackson, Fernandes, and Samick producing many different versions with slight changes from the original, such as inverting the shape of the headstock.

Hard rock and heavy metal players have been the leaders in adopting iconoclastic designs and new finishes, such as checkerboarding or modernistic black and white "graphics," on their Flying Vs and Explorers. Unusual, eye-catching designs with outrageous shapes and finishes have become the norm for heavy metal guitars. The favored shape of a typical heavy metal guitar would most likely be based on the Explorer or the Flying V, with the obligatory high-output humbucker pickups and locking tremolo unit. Heavy metal guitars are built more for speed than for tone. Speedy, precise fretting and moving quickly up and down the neck require narrow necks and low action—in which the strings are closer to the frets—to facilitate the fast runs and frenetic speed of solo lead lines.

The development of the archetypal heavy metal guitar owed a lot to the input of the virtuoso guitarists who played a much larger role in design and manufacturing in the 1980s. On one hand this was the result of their immense stature among purchasers of guitars, and on the other it reflected the growth of custom guitar builders. The custom shop came of age in the 1970s and 1980s when the large manufacturers were suffering from a reputation for poor quality. Custom builders usually had all the right credentials as craftsmen, and their practice of making one guitar at a time, usually for a star player, was the ideal counterbalance to perceptions of shoddy mass-produced guitars made by companies who did not care about the music.

The typical development path of the custom builders can be illustrated by Bernie Chavez Rico, who started in the late 1960s. The beautiful and exotic woods he used for the bodies and the advanced electronics attracted many famous customers, and as B.C. Rich Guitars he moved into larger production runs, using some foreign-made parts. B.C. Rich guitars are known today for radical shapes and are closely identified with the celebrated players who endorse them. One of the most memorable sights of 1980s pop music was the outrageous C. C. DeVille of Poison strutting around the stage with his fluorescent B.C. Rich guitar.

Guitarists had always customized their instruments; as soon as they got hold of Fender's Telecaster, they took the metal cover off the bridge so that they could rest their palms on the bridge to mute the sound. But in the era of the heavy metal god one man's customizing could become the blueprint for a new mass-produced guitar. Because numerous small companies made component parts of the electric guitar system, players could create their own guitars from parts. For example, a wide range of pickups could be purchased from companies like DiMarzio and Seymour Duncan. One of the most famous guitars of the 1980s, Eric Clapton's "Blackie" Stratocaster, was constructed by the player out of several different instruments. When Eddie Van Halen attached a Gibson PAF humbucker pickup to a guitar made up of Stratocaster parts, he broke a long-established format and established some important precedents. This was his "Frankenstein" (or Frankenstrat) guitar, which he used on the first albums to create important songs like "Eruption" and "Jamie's Cryin'" (see plate 10). It was the pioneer of the next generation of Stratocaster clones—the Superstrats—which drew on a wide range of non-Fender technology to enhance the potential of this instrument, such as Seymour Duncan pickups and Floyd Rose tremolo units.

Locking tremolo units were very important to the hard rock and heavy metal players, who loved to work the "whammy bar" to excess. Jimi Hendrix had demonstrated what could be done with the Stratocaster version, but techniques like "divebombing," which used violent tremolo, required a unit that could stand up to a lot of use without pulling the strings out of tune. The Kramer guitar company forged a very important relationship with Eddie Van Halen, and when the company's Rockinger tremolo was found wanting by its premier customer, Kramer adopted a double-locking unit devised by Floyd Rose, a guitar player and machinist who was then making them in his basement. Van Halen found that this design did not pull the strings out of tune, and Floyd Rose was soon out of the basement and into the big business of manufacturing and licensing his innovation. The patronage of a heavy metal god like Van Halen was the catalyst to turn a custom shop operation into a nationally known guitar manufacturer. Van Halen's Frankenstrat was turned into a line of Baretta models by Kramer, and this partnership put the company firmly on the map.

the heavy metal show

The dramatic changes in the way the electric guitar looked was partly the result of the manufacturers' manipulation of high-tech symbolism, and partly the rise of spectacle in 1970s popular music. The electric guitar had a vital symbolic role to play in this spectacle. It was not enough to have a original sound; one needed an original instrument too. Musicians were aware of the signals that the look of their instruments sent to the audience. The story is told of Eddie Van Halen's band-mates complaining that his Gibson ES-335 with its traditional design did not look "rock" enough for the image of the group.

The hard rock and lite metal players needed dramatically new instruments to reflect their exaggerated stage shows and their unworldly star personas. In the 1970s guitars were called axes and the virtuoso players axmen. Gene Simmons of Kiss had a guitar made in the form of a medieval ax—a weapon rather than an instrument—and the Ovation Breadwinner looked like an ax with its curved, beveled base. The ax denotes power and force, and it is usually in the hands of a man. While the members of Kiss were not virtuoso players, they introduced a new concept of the electric guitar to millions of new fans, one that dealt less with the sound and more with what it represented. Their appearance in magazines devoted to guitar players outraged the traditionalists (who were still devoted to blues rock), but as one publication pointed out, this had been the initial reaction to Jimi Hendrix in the 1960s! Kiss claimed to be first in many important categories of guitar-based spectacle: first to have a smoking guitar, first to have a guitar that shot rockets, and first to have a guitar covered in lights that flickered like a movie marquee. Ace Frehley of Kiss commented that the point of the showmanship was to make the band "larger than life—we suddenly became superheroes."[11]

The impact of Kiss was felt the most by the very young, and it was an experience that leaned as much on the look as on the sound of the electric guitar. As one fan remembered about those heady years in the 1970s: "I would stand in a Paul Stanley stance holding a baseball bat like a guitar, then the guitar intro builds and the drums are like fireworks and cannon explosions. To a seven year old this is what rock'n'roll is all about!"[12]

Everything about Kiss was fantastic: their outfits, their guitars, their stage shows, and their success in filling stadiums and selling records. The imaginary personas they created were larger than life and perfectly suited the rock concert that had taken on the aura of being a unique, otherworldly event with visitations rather than presentations of performers. In lite metal shows the leather pants, the lights, the smoke bombs, and the posturing with rampant guitars were all accepted as part of the performance. Many strains of speed and thrash metal avoided conspicuous stage shows, but other subgenres, such as death metal, had carefully thought-out shows that focused on gloomy, gothic themes. Although parents and legislators were horrified at the dark themes explored by some heavy metal acts, the players were operating in the same forbidden zones as their 1950s predecessors; as Ozzy Osbourne sang, "They say I worship the devil / But I'm just a rock'n'roll rebel."

While lite metal and hard rock were favorably received by the mainstream music industry, many metal bands existed in happy exile from it—untainted by commercialism and selling out. The lack of radio play forced heavy metal into the arena of performance, often highly theatrical entertainment that blended visual and aural stimuli into an exploration of the fantasy underpinning the songs. The primacy of the stage show was a celebration of both the music and its followers, who came together in an elaborate ritual of fantasy and power. Robert Walser describes it as a dialectic between the forces of freedom and control—the control of this powerful generator of sounds, which can also give the player (and his audience) the transcendental experience of liberation and freedom. What Walser hears in the obligatory heavy metal solos are rhetorical outbursts that empower and liberate those who hear them. The virtuoso heavy metal player manipulates shared musical codes that reach his audience as a powerful emotional event. The volume of the music translates into a shared enjoyment of power between musician and audience, and in Walser's analysis this musical articulation of raw power brings a transcendental experience to the young men in the audience.

The attention of the crowd was usually directed at the vocalist, the frontman and spokesperson, but the climax of the song was invariably a guitar solo. The lead guitarist not only had to play well; he had to dramatize his playing as part of the performance: he would bend over his instrument, his face a model of concentration and exertion as the fans looked on in awe and respect, many of them playing air guitar along to the music. As Harris Berger points out, the metal guitarist uses the stage show to make a display of emotional engagement to the music. The guitar assumes importance as the symbol of supernatural power, youthful rebellion, and sexual prowess. Lite metal players probably took the phallic powers of the electric guitar to their highest level in their celebration of masculine hedonism. The great stadium shows of the 1970s and 1980s covered many different genres of music and used spectacle in different ways (see plate 11). The only common thread is that unprecedented levels of amplification were required and the electric guitar had center stage. David Bowie's imaginary Ziggy Stardust conjures up a delightful image of extraterrestrial music and gender, but first and last we are given this vital piece of information about him: "Ziggy plays guitar."

How much the image of the professional guitar player changed during three decades of rock 'n' roll can be gauged by comparing two images, one from the late 1940s and one from the late 1970s (see plates 12 and 13). We see two guitar players: one is Les Paul and the other is Angus Young, lead guitarist from the hard rock band AC/DC. One player is formal and stiff and holds the instrument in a horizontal position for ease of playing. He is dressed in a jacket and tie and has the look of a professional as he smiles at the viewer. The other guitar is nearly vertical and is played by what appears to be a naughty child—disheveled, frantic, with tongue pointing out. One image is restrained and sedate, the other anarchic and subversive. The bolt of lightning coming from Young's foot underlines the power at his disposal and hints at the transcendence of the music.

Sources

Berger, Harris M. *Metal, Rock, and Jazz: Perception and the Phenomenology of Musical Experience*. Hanover, N.H.: University Press of New England, 1999.

Walser, Robert. *Running with the Devil: Power, Gender, and Madness in Heavy Metal Music*. Hanover, N.H.: Wesleyan University Press, 1993.

Weinstein, Deena. *Heavy Metal: A Cultural Sociology*. New York: Lexington, 1991.

Notes

1. Walser, *Running with the Devil*, p. 68.

2. Weinstein, *Heavy Metal*, p. 122; Robert Duncan, *The Noise* (New York: Ticknor and Fields, 1984), p. 47.

3. Van Halen quote: "How the Other Half Rocks," in *Rock Eras 1954–1984*, ed. Jim Curtiss (Bowling Green: Popular Press, 1987), p. 288, citing Jas Obrecht, *Masters of Heavy Metal* (New York: Quill, 1984).

4. Walser, *Running with the Devil*, p. 42.

5. *Guitar*, April 1999, pp. 33–38.

6. Ozzy Osbourne, interview in Penelope Spheeris's documentary *The Decline of Western Civilization Part II: The Metal Years* (1988).

7. Martin Popoff, *The Collector's Guide to Heavy Metal* (Burlington, Ont.: Collectors Guide, 1997), p. 281.

8. Advertisment in *Guitar Player*, April 1972.

9. Via Rob Timmons, Fernandes Guitars International, www.fernandesguitars.com, 9 July 2003.

10. Dean Zelinsky, interview at www.deanguitars.com, 19 June 2001.

11. Ace Frehley, interview in *GuitarOne*, February 2002, pp. 72–73.

12. Beam collection of oral histories of musicians, University of Alabama at Birmingham.

Further Reading

Reynolds, Simon, and Joy Press. *The Sex Revolts: Gender, Rebellion, and Rock'n'Roll.* Cambridge: Harvard University Press, 1995.

9

women guitarists

gender issues in alternative rock

John Strohm

Can women play guitar? Having formed my views in the sheltered environment of the 1980s Boston underground music scene, I have long unquestioningly believed that women are equal to men on the rock 'n' roll stage. These views took shape while I played in the alternative rock trio the Blake Babies, alongside drummer Freda Love and guitarist–lead vocalist Juliana Hatfield. Although we rarely discussed the role of women in rock or sexual politics in general, I assumed that my band-mates subscribed to the same views as I did. It came as a great surprise when, after the group broke up in the early 1990s and Hatfield embarked on a successful solo career, she began to express somewhat inflammatory views on the subject of female electric guitarists. Her comment in the October 1992 issue of *Request* magazine is typical: "I can't think of one woman in rock doing anything great, but I don't think it's because people don't let them. Men just play better than women. I think maybe it's genetic. (Perhaps) we have one hormone missing."

On a recent reunion tour with the Blake Babies, I had the opportunity to discuss these statements with Hatfield to see if she still felt the same way. Faced with the challenge of writing something meaningful about women guitar players, I went straight to the controversial source. The strength and conviction of her recent statements, made nearly a decade after the one in *Request*, took me by

The Blake Babies: Juliana Hatfield, John Strohm, and Freda Love. Photo by Gary Smith.

surprise. She not only held the same views, but she had sharpened them to a single, incontestable point, arguing, "Yes, I still believe women can't play electric guitar nearly as well as men, and they never will. Just think about it: how many women are there who have actually been innovative, technically accomplished lead guitarists? None! There aren't any. All of the great, original, innovative electric guitarists have been men. There is no point in even discussing it." As a respected electric guitarist who is widely considered to be one of a handful of innovative women in the field, Hatfield seemed to deny her own contribution to the culture of the electric guitar, along with those of her female peers.

While I instinctively disagreed with her, I had to admit that she presented a strong argument. I could think of many technically proficient female electric guitarists, such as Bonnie Raitt, Lita Ford, Sylvia Juncosa, and Louise Post, but none of them had significantly changed the way the instrument is played. All of the great innovators of the electric guitar, from Charlie Christian to Chuck Berry, from Jimi Hendrix to Eddie Van Halen, have been men. Sitting there listening to Juliana, I couldn't think of a good counterargument, but I knew that women had significantly contributed to the development of the instrument, albeit in more subtle ways. The story of women electric guitarists has been a constant struggle to redefine a musical instrument that, somewhere in its evolution, became identified as a signifier of masculinity. In addition to breaking gender barriers, women guitarists have also contributed to the way the instrument is played. The redefinition of the electric guitar in terms of the female musician represents the success on one of the final frontiers of the women's movement: the deconstruction and reconstruction of macho culture and its signifiers.

With its soft curves and sharp angles, the solidbody electric guitar cuts a distinctly feminine figure. Although the object may resemble the female form, its central role in rock music and culture is as a signifier of masculine power and implied sexual prowess. The lightning-fast, screaming lead lines played by the guitar hero testify to his immense machismo and form the foundation of his mythical status in rock culture. Face contorted in quasi-orgasmic bliss, Gibson Les Paul at crotch level pointed toward the ceiling, the guitar hero rips blues licks at mind-bending speed: this is the quintessence of rock. So where do the women who play the electric guitar fit into this picture? Traditionally, at least in the world of hard rock and heavy metal, women have been relegated to subservient roles of groupie, muse, or care giver. While female acoustic guitarists have long been accepted and frontwomen in rock bands have become commonplace, women electric guitarists have struggled to overcome prejudices derived from the macho image of the electric guitar.

Before the rock era the solidbody electric guitar was not such a loaded signifier of masculinity. Early 1950s publicity photos of husband and wife duo Les Paul and Mary Ford depict them posing with their Gibson electrics in various comfortable domestic settings, smiling and plucking away while living the postwar suburban dream. The seeming awkwardness (to modern eyes) of their electric guitars, the very symbol of macho rock rebellion, in such a setting calls attention to the enormous cultural change the instrument represents. With the advent of rock the electric guitar has come to symbolize the rejection of the very cultural values Les and Mary espoused: domesticity, monogamy, safety, and security.

These values also happen to be culturally associated with femininity, so in a sense rock was a rejection of a traditionally feminine value system. This might explain the commonly held feeling that electric guitars look strange on girls. With the electric guitar as a symbol of masculinity, a woman strapping on and plugging in a Strat or Les Paul became tantamount to a woman wielding a chain saw or driving an eighteen-wheeler; it offended our conventional notion of femininity.

A woman playing the electric guitar challenges gender stereotypes, so in one way even the most conventional women players have been innovative. Some of the best-known female players, such as slide blues master Bonnie Raitt, jazz virtuoso Emily Remler, and heavy metal shredder Lita Ford, have excelled at styles pioneered by men, pushing parameters only in the sense that their playing within the established styles is outstanding. But because these women chose to play in the mainstream, they are less controversial than the innovators who rejected conventional ways of playing the instrument. Following the mid 1970s punk movement a series of maverick women have not only challenged the instrument's macho image but also opened new possibilities for the electric guitar as a means of personal expression. While some critics have charged that women lack the physical and mental characteristics necessary to play lead guitar, many women players in recent years have rejected conventional styles in favor of their own unique, personal approach.

Although the electric guitar didn't become a full-fledged cultural phenomenon until the 1950s, a series of pioneering guitar players in the 1930s and 1940s opened up new possibilities for the invention. While female acoustic guitarists were common throughout the nineteenth century, the increasing size of the instrument contributed to the decline of women players in the early twentieth century. By the 1930s female guitarists were a novelty in an increasingly male-dominated field. Two rare examples of innovative female players in the prerock era are Memphis Minnie and Sister Rosetta Tharpe. Both originally acoustic guitarists, they switched to the electric initially for increased volume at live performances. As a Chicago blues contemporary of Big Bill Broonzy and Muddy Waters, Memphis Minnie helped to define a style of playing that exploited the unique sound of the electric guitar.

The virtual absence of female electric guitarists in the rock scene of the 1950s had more to do with women's place in society than with their physical or mental limitations. The female electric guitarists in Bo Diddley's touring act, such as Lady Bo and the Duchess, were featured as a novelty rather than for their unique talent, and they never recorded on Diddley's records, although his maraca-shaking sidekick Jerome Arnold did. Women such as Lorrie Collins and Wanda Jackson emerged as rockabilly frontwomen but always appeared before all-male bands with male guitarists handling the difficult lead work. Save for the occasional novelty instrumentalist, such as Kathy Marshall, the thirteen-year-old "queen of the surf guitar," the early rock business tended to reflect American society of the 1950s, with females as passive participants relegated to the sidelines of fandom and domestic support.

A significant breakthrough occurred with the cultural changes precipitated by the publication of Betty Friedan's *Feminine Mystique* in 1963. The women's movement questioned the traditional role of mother and homemaker. Changes in sixties pop-music culture underscored these challenges. The move from girl groups of the early sixties to the psychedelic rock scene late in the decade recast the passive women singers as empowered frontwomen with a measure of creative control in their male-dominated bands. For example, male impresarios such as Phil Spector and Berry Gordy strictly controlled girl groups in the early sixties. The female singers in groups such as the Ronettes and the Supremes performed songs composed mainly by men and were backed up by male musicians, with a few notable exceptions later in the decade such as the electric bassist Carol Kaye. But a few years later performers such as Janis Joplin of Big Brother and the Holding Company or Grace Slick of the Jefferson Airplane took a more active role in the creation of their music, even though the hippy scene as a whole still relegated women to subservient roles.

The birth of the women's movement also coincided with the immensely popular folk revival of the 1960s. This rediscovery of folk music empowered women because it sent the message that they could write and perform their music without male support and they could detail their distinctly feminine perspective in daring, confessional lyrics. Extraordinarily popular performers such as Odetta, Joan Baez, and Joni Mitchell commanded respect from their male contemporaries and provided strong role models for aspiring women guitarists. The expansion of the women's movement also helped to spread a "can do" attitude among young women that led directly to their first forays into rock music in the late 1960s and early 1970s. Feminist writers such as Patricia Kennealy-Morrison loudly questioned the virtual absence of women in rock. In a 1970 article provocatively entitled "Rock around the Cock," she bravely articulated her own electric-guitar-playing fantasy: "I want to get up onstage at the Fillmore East, wearing a black leather jumpsuit and a silver-plated Telecaster, grab the mike, sneer at the audience, 'you PIGS,' then get off forty-five minutes of the undisputedly finest rock guitar ever heard anywhere."[1]

Despite the increasing influence of women in rock in the 1960s, few role models existed for aspiring female electric guitarists. While the rare female studio musicians in rock 'n' roll, like Carol Kaye, largely operated behind the scenes (thus minimally challenging male domination), other electric guitarists, such as Kathy Marshall, assumed the tried-and-true role of novelty act. The reason women were not taking to the electric guitar lies in the technology itself. The amplified electric guitar was perceived as an instrument of great power, and this was the main reason that its playing in public was restricted to males. Electric guitars were seen as a man's preserve; thus the fraternity of professional guitar players was closed to women as equals. Also the complex equipment that made up the technological system required a certain amount of technical savvy, a traditionally male attribute. While the folkie acoustic players flourished, the electric guitar remained a boys' club. This reinforced the idea that electric guitars were for boys and acoustic guitars were for girls.

In an odd twist in light of the burgeoning women's movement, early 1970s "glam" rockers such as David Bowie and the New York Dolls began to experiment with gender reversal. Men sporting makeup and platform shoes on stage opened the possibility for women to try out traditionally male roles in popular music, and this provided the foundation for the first real breakthroughs for female electric guitarists. The first popular all-female rock band, Fanny, emerged in 1970 and revealed the commercial possibilities for the format. The musicians in Fanny played the traditionally masculine electric guitar, electric bass, and drum set and succeeded on several levels: there was the well-established novelty of women playing male roles, and there was also the fetish appeal of the "bad girl" image that presented female players as toughs (thus also fitting into the traditional macho image). Fanny revealed a new marketable niche, and in the market-driven music business of the 1970s, several acts sought to exploit that opportunity.

Suzi Quatro is a complex figure, because she is both a blatantly commercial pop artist and a giant in the history of female electric guitarists. She began her career as a teenage member of the obscure 1960s Detroit all-girl garage band the Pleasure Seekers. In her subsequent solo career, she performed songs—assembly-line pop usually composed by men—but her image as the leather-clad, swaggering, bass-playing bad girl influenced a generation of female guitarists and bassists. Debbie Smith of the 1990s British rock band Echobelly recalls of Quatro: "She was dressed head to foot in black leather and she was the lead singer with this whacking great bass. That was it. I thought: right, I want to play the bass guitar."[2] Quatro did not achieve nearly the popularity in the United States that she did in England, but she did manage to influence another tough-girl rock pioneer, Joan Jett.

Modeling herself after Quatro, California teenager Jett joined the all-female teenage Runaways in 1976. Assembled by Kim Fowley, LA rock scenester and former member of the Byrds, the Runaways constructed their own updated version of the tough-girl image, which became the blueprint for female hard rock bands for decades. Originally intended to be an outlet for Fowley's songwriting, the Runaways quickly began composing their own songs and crafting their own image. Although the Runaways played their instruments well, their main selling point was the spectacle of attractive teenage girls playing aggressive rock music. They willingly traded on their sex appeal in their double-entendre lyrics and revealing outfits, reducing themselves to sensationalism despite their considerable talent. While men seem to have responded to the fantasy element of girls taking on the traditionally masculine role of heavy rock musicians, the band also connected with a female fan base. Live recordings reveal the shrill screaming of large groups of empowered adolescent girls. Although the Runaways never achieved much success in the United States, two members of the band, Jett and Lita Ford, did make their mark on popular culture.

In her solo career Joan Jett crafted an image that distilled all the elements of her former bad girl into an extremely marketable product. Her debut single served as her manifesto when she sang, "I don't give a shit about my bad reputation." Her two massive breakthrough singles, "I Love Rock N' Roll" and "Crimson and Clover," were covers of earlier pop classics by the Arrows and Tommy James, respectively. Her main selling point was her rebellious image of a sneering, leather-clad bad girl playing an electric guitar. She exuded a "macha"[3] that distinguishes the can-do attitude of a certain class of female rock musicians who seek to "out-macho" the men in the masculinity contest that is hard rock. In the seventies this macha helped to create a market niche and placed the female guitar player in the public eye. In her update of Suzi Quatro's image, Jett announced to a nation of girls that playing electric guitar was cool. Her impact on both the music industry and the next generation of female guitarists is immeasurable.

punk

Jett's breakthrough owed as much to the doors kicked open by the punk explosion of the 1970s as it did to her catchy songs. Jett's recrafted image incorporated punk elements that had now become acceptable in pop music. Punk was a reaction against the bland, formulaic *business* rock had become, and for Jett and many other female musicians, it provided the greatest breakdown of the gender rules in the history of rock. Few women, with notable exceptions such as Gaye

Advert (bassist for the early British punk group the Adverts), actually played instruments in punk bands, but punk demystified rock, announcing that one did not need to be a virtuoso musician or a man to get on stage or make records. One of the central tenets of the punk ethos was the rejection of the rock star and the guitar god. This back-to-basics movement argued that much of the emotion of the music and its power to communicate had been lost in flashy lead guitar playing.

A famous illustration in *Sniffin' Glue,* a punk fanzine from 1976, showed three primitive chord diagrams (for an A, an E, and a G) and then told readers, "NOW FORM A BAND." Lifting the barrier of virtuosity opened up guitar rock to the untutored masses. Phil Caivano of Monster Magnet remembered his anxieties about his skill as a guitar player and the awesome responsibility of reaching the standards set by the guitar gods, "But punk rock gave me the nerve to get up and do it." Bands like the Ramones encouraged others to get on stage: "You just gotta play boys. . . . come out of your basement and play. That's what we did." Lucy Toothpaste reported, "Boy bands were getting up on stage who couldn't play a note, so it was easy for girls who could not play a note to get up on stage as well."[4]

Although most women in early punk bands served as noninstrumentalist lead vocalists, they differed dramatically from earlier frontwomen. American Patti Smith and Londoner Siouxsie Sioux rebelled against feminine gender roles and the expectations of the lead singer. They adopted a tough, unladylike pose that borrowed more from the macho swagger of sixties garage bands than from the calculated bad-girl image of bands like the Runaways. They went beyond the leather outfits to the bondage gear of Sioux and the straight-from-the-gutter androgyny of Smith. They articulated a female rage that surpassed the anger of the women's movement of the sixties. Smith's version of "Hey Joe" has a woman terrorist killing her unfaithful lover, and Sioux's "Suburban Relapse" describes a woman revolting against the domestic regime. Their songs see marriage as a prison ("Happy House") and stress the emptiness of consumer culture for the married woman ("Spend Spend Spend"). By the time the Sex Pistols registered on American cultural radar, female-fronted punk bands such as the Avengers, the Vice Squad, and X were common in both America and the United Kingdom.

Patti Smith and Siouxsie Sioux were the first role models for women in punk. The British group the Adverts emerged in 1977 with the seminal single "Gary Gilmore's Eyes," setting a precedent for women instrumentalists in punk. Shortly thereafter Londoner Poly Styrene formed a band with a male rhythm section and a sixteen-year-old sax player named Lora Logic. X-Ray Spex featured two aggressive women fronting the band and delivering songs with lyrics like "Some people say that little girls should be seen and not heard / But I think Oh bondage, up yours!" As bands like the Adverts and X-Ray Specs gave voice to the rage and ennui of British youth, London businessman and Sex Pistols impresario Malcolm McLaren saw another commercial opportunity. He groomed expatriate American music journalist, singer, and rhythm guitarist Chrissie Hynde to front an all-male punk band called the Love Boys. In addition to working in McLaren's shop, Hynde played in several punk groups and, on the strength of her demo tape, managed to raise enough money to form her own band, the Pretenders. Hynde played the leading role as songwriter, vocalist, and rhythm guitarist.

Like Patti Smith, Hynde crafted an image for herself that drew on several male archetypes of rock 'n' roll rebellion, aping the swagger and low-slung Fender Telecaster of vintage Keith Richards and the sneer and snotty delivery of protopunk Iggy Pop. She maintained a keen pop sensibility, her songs conveying toughness but with an underlying vulnerability. Here was a female electric guitarist who had something new to say to women, and her importance as a role model should not be underestimated. Nina Gordon, rhythm guitarist and co–lead vocalist of nineties American rock band Veruca Salt, recalled: "I remember being completely floored. I heard that voice, saw that picture of Chrissie Hynde with her guitar . . . how did she know how to do that?"[5] The Pretenders achieved huge commercial success in America and in Europe, so she was able to have the same kind of massive impact as Joan Jett had. Hynde's influence was arguably greater because she distinguished herself as an outstanding songwriter. She clearly established that a woman could lead a rock group to great success without trading on her sexuality or pandering to pop trends or relying on preestablished female roles.

As the punk movement expanded in the wake of the Sex Pistols' international success, women began to take punk music in unexpected new directions. The all-female British band the Slits completely shunned conventional instrumental technique and opted for a primitive approach that conveyed pure, raw emotion. The cover of their 1979 debut album *Cut* depicts them stripped to the waist, clad only in loincloths, and covered head to toe in mud. Like many of their powerfully feminist songs, this photograph challenged the notion of the female form as a commodity, which until then had been the driving force behind all-female rock groups. This sort of confrontational, image-driven social commentary became a key element of many female rock bands in the 1980s and 1990s. The Slits also broke new ground for female musicians by boldly confronting previously taboo gender issues. Their song "Typical Girls" rejected consumer society and its predetermined obsessions with body shape, cleanliness, and romance. Their music reflected a changing definition of femininity, and the fifties ideal of the domesticated woman became outmoded as a new musical culture permeated all corners of the Western World. The Slits's unique brand of idiot-savant reggae punk provided another example for female guitarists. They announced that literally anyone could do it: thin or fat, ugly or beautiful, technically proficient or absolute beginner. Along with their punk peers, they also proclaimed that virtually no subject remained out-of-bounds for their songs.

The Raincoats, like the Slits, formed at the height of the British punk movement and celebrated the right of untrained guitar players to make music in public. The "do it yourself" (DIY) ethic was a central tenet of the punk philosophy, and Raincoats founders Gina Birch and Ana Da Silva took it to heart. They bought guitars and formed their band before they learned how to play. In a reaction against the bloated, overwrought rock concerts of the day, punk blurred the line between performer and audience. The Raincoats had attended punk shows and noticed that "people you'd see in the audience one week were up on stage the next."[6] Their empowerment in music reflected their new status as women, and they gleefully reminded the audience that they were "no one's little girl." Their lack of musicianship freed them of clichés, and as a result they created some of the most unique music of the punk era, which seemed to spring directly from the id, unencumbered by the collective ego of rock tradition. The Slits and the Raincoats, along with the all-female Swiss band Liliput, represent the initial emergence of female electric guitarists who owed nothing to the earlier innovations of male players. Suddenly anyone could pick up the guitar, learn a few chords, and make a perfectly valid statement of her own within this new, unrestricted world of guitar players.

By the late 1970s the New York punk scene began to merge with the experimental Lower East Side jazz community to forge a new style of abrasive, confrontational antipop music known as "no wave." As its name indicates, no wave challenged the more commercial genre, called new wave, that emerged in the wake of punk. The no wave bands played a primal yet sophisticated style of music that took punk rock a step further. Within the context of electric guitar, electric bass, and drums they introduced expressive free-jazz elements such as skronking horns and abrasive, metallic percussion. Lydia Lunch fronted the no wave band Teenage Jesus and the Jerks, and her diabolical slide guitar style revealed new expressive possibilities for the electric guitar. Expressed through her vocals and guitar playing, her rage surpassed all of her punk predecessors. The fact that she lacked conventional proficiency on her instrument made little difference: in the terrifying musical landscape of no wave, screeches and wails replaced chords and scales as the building blocks of songs. A generation of bands such as Sonic Youth, Big Black, and the Swans took their cues directly from the groundbreaking music of Teenage Jesus and no wave.

The creative spirit of the late 1970s downtown New York scene provided opportunities for several talented, innovative female guitarists. The Cramps' schlock-horror-themed punk rockabilly relied heavily upon guitarist "Poison Ivy Rorschach"'s aggressive, fifties-influenced sound. Her "ice queen" stage persona presented an image of an empowered woman who could knowingly flaunt her sexuality in a way that transcended exploitation. She also proved that women guitarists in the punk scene could play traditionally rooted bass guitar styles (such as funk and soul) as well as men. Talking Heads bassist Tina Weymouth opted for a more androgynous look, downplaying her gender. Her playing pushed the parameters of punk and new wave by incorporating elements of funk and Afro-pop. Sonic Youth bassist Kim Gordon emerged from this scene and boldly affirmed her feminist views with lyrics such as "support the power of women / use the power of men."

Taken as a whole, the punk and new wave bands formed in the late 1970s did a lot more than provide role models for women who wanted to make music professionally. They provided a feminist perspective for rock 'n' roll songs that for over two decades had represented masculine viewpoints. They brought a new honesty in depicting relationships from a woman's point of view. But equally important, they gave women a new confidence in dealing with the male world of musicians and guitar technology. As Louise Post of Veruca Salt argued, "the feeling of picking up a guitar and just making noise is a really powerful one . . . why should men have a monopoly of loud music?"[7]

A handful of innovative female guitarists in the international punk community revealed the possibilities for women in the postpunk generation of the 1980s. While the British postpunk style moved away from the electric guitar in favor of a slick, synthesized sound, the hardcore element of American punk music stripped the original sound down to its essential elements and greatly increased the tempo. While the East Coast hardcore scene developed into a macho, subtly homoerotic boys club that largely excluded women, the West Coast scene welcomed female musicians. Leading Los Angeles hardcore band Black Flag featured female bassist Kira in their lineups, and singers such as Exene Cervenka from X and Penelope Houston from the Avengers provided a much-needed women's perspective to this hyperaggressive genre. The early hardcore scene laid the groundwork for the massive infiltration of women guitarists in the late 1980s and early 1990s, and it also facilitated the first real commercial breakthroughs for the all-female rock groups the Go-Go's and the Bangles.

girl groups in the mainstream

Belinda Carlisle, lead singer of the Go-Go's, began as the drummer for the seminal LA hardcore punk band the Germs, which also featured female bassist Lorna Doom and a classic debut album, *(GI)*, produced by Joan Jett. Carlisle formed the Go-Go's in 1978 with guitarists Jane Wiedlin and Charlotte Caffey, both veterans of the LA punk and new wave scene. They added the aggressive, highly competent rhythm section of drummer Gina Schock and bassist Kathy Valentine. The Go-Go's supercharged sixties mix of surf music and girl-group vocalizing created a sound that connected perfectly with the pop music market. The commercial success of their first album, *Beauty and the Beat,* proved that women playing electric guitars in a pop context, free of the macho trappings of their "bad girl" predecessors, could sell millions of records. While the Go-Go's never repeated the commercial or creative success of their debut album in their five-year career, they certainly influenced the next generation of women electric guitarists. According to Veruca Salt guitarist Nina Gordon, "I spent hour after hour with headphones listening to it [*Beauty and the Beat*], singing along and fantasizing about being one of them." According to Belinda Carlisle, the Go-Go's were successful because "all the boys wanted to sleep with us, and all the girls wanted to *be* us."[8]

The Bangles began as the Bangs in 1981, around the time of the Go-Go's initial commercial success. This all-female band became part of the local 1960s revivalist Paisley Underground scene, along with fellow LA bands the Rain Parade, the Three O'clock, and the Dream Syndicate, featuring bassist and occasional lead vocalist Kendra Smith. Although the Paisley Underground existed apart from the thriving hardcore scene, both relied on the same underground network of fanzines, independent labels, and college radio stations to promote their music. The Bangles borrowed the jangly guitar sound and tight harmonies of the Byrds and blended them with the sassy, empowered attitude of the original sixties girl groups. While the rich, serendipitous blend of women's voices provided their distinctive sound, guitarists Vicki Peterson and Susanna Hoffs successfully reproduced the classic twang of the sixties with a slick, modern sound.

The evolution of the Bangles from critics' darlings to MTV-friendly hit machine mirrored the rapidly changing LA rock scene in the 1980s. As the Bangles became assimilated into the mainstream landscape, highly fashionable but disappearing as instrumentalists on their own recordings, women were rarely seen in the hard rock and metal scenes. While makeup-wearing, cross-dressing male rockers became ultramacho caricatures of 1970s glam rock icons, women found themselves in regressive roles such as groupies and care givers. The balance of power, at least in mainstream music, had shifted back to clichéd macho-male rock stars. The rare all-female bands in the metal scene, such as LA's Vixen, reverted to the macha of earlier bands such as the Runaways and Girlschool, but with a new emphasis on visual presentation that reflected the MTV-driven market.

The 1980s metal scene largely relegated women to traditional, subservient roles, but certain exceptional lead guitarists found success within it, such as two-hand tapping virtuoso Jennifer Batten and former Runaway Lita Ford. Batten emerged from the LA metal scene to score an enviable gig as Michael Jackson's lead guitarist in his late-eighties touring lineup. She fought gender assumptions and stereotypes throughout these tours: "It's a shock for some people to see a woman playing [electric] guitar. All over the world [on the Michael Jackson tour] people would ask me if I was a man or a woman. Just because I played guitar, they assumed I was a guy. I'd stand there with my blonde hair, red lipstick and caked-on stage makeup thinking, 'thank you, Poison! Thank you, Cinderella! You've confused the children of the world.'"[9] Batten eventually became the first female instructor at the prestigious Guitar Institute of Technology and a columnist for *Guitar Player* magazine, opening doors for women in the male-dominated world of electric guitar education.

Following the breakup of the Runaways and the huge success of Joan Jett's first two albums, Lita Ford emerged as a solo artist and a true heavy metal purist. Influenced by the classic metal of Ritchie Blackmore and Michael Shenker, Ford's fluid lead guitar style favored "feel," the technical flash that dominated eighties metal. A rare but powerfully outspoken female presence on the metal scene, Ford actively encouraged girls to make the necessary sacrifices to play the electric guitar. "Many women don't play because they think it's going to take away from their femininity because they can't have long fingernails. But if you really want to be a guitar player, you have to sacrifice something. My nails are gone, but I don't care."[10] Ford's success continued throughout the heavy metal boom of the 1980s, eventually dying out in the 1990s as this type of hard rock fell out of favor.

The reactionary mainstream rock and pop scenes of the late 1980s made it difficult for women to break through, but a new urban underground music culture began to emerge that empowered female musicians to an unprecedented degree. In the early punk scene female musicians in predominantly male bands had been something of a novelty, but suddenly no one seemed to notice when an underground rock band featured a female bassist or guitarist. In postpunk pioneers like Sonic Youth, the Feelies, and Bush Tetras, and late-eighties newcomers such as Superchunk, the Pixies, and Galaxie 500, women guitarists and bassists commanded the same respect and earned the same degree of credit as their male band-mates.

Along with increasing female membership in male-dominated bands, all-female or predominantly female bands became more common and more stylistically diverse in the late eighties and early nineties. The underground rock scene (which was eventually christened "alternative" rock in a successful attempt by the music industry to create a marketing niche) initially rebelled against the crass aspects of mainstream music. Bands shunned such pretentious conventions as the stage outfits and choreographed moves that dominated the late-eighties mainstream. Inspired by the DIY nature of the punk movement, they chose to further blur the lines between band and audience, stressing the point that literally anyone could make music. Band lineups that mixed race and gender contributed to a greater social awareness within the rock community, calling into question rock's sexist past. Women felt free to express themselves as never before, and an amazing variety of female-fronted mixed-gender or all-girl bands emerged.

from riot girls to lilith fair

A guitarist and drummer since my teenage years, I began playing in punk bands in my hometown of Bloomington, Indiana, in the early 1980s, but my musical career really began in Boston a few years later with the formation of the Blake Babies. Although I shared the stage and studio with Freda Love and Juliana Hatfield, I don't recall ever considering my band-mates to possess any musical disadvantage because of their sex. Because I'd grown up playing in punk bands, I was very comfortable with the idea of playing with women, despite the fact that this was a radical idea in mainstream rock culture. Boston was arguably the most progressive rock scene in the country in terms of sexual integration, with countless bands such as the Pixies, the Throwing Muses, Galaxie 500, Christmas, and Salem 66 sporting mixed-gender or all-female lineups.

While my band-mates and I may have felt comfortable in and around Boston, we certainly encountered resistance and ridicule on the road, especially in the South. Club owners and promoters frequently encouraged Love and Hatfield to wear dresses and makeup, reasoning that they would benefit from exploiting their beauty on stage. (The style of the day for alternative rock acts was unpretentious to the extreme—jeans and T-shirts, no makeup. Our credo was to wear on stage whatever outfit we happened to be wearing that day.) The antifashion of the American underground was a very conscious reaction against the flamboyant presentation of mainstream, corporate music. Although men dressing down received little or no flack from the public, women dressing down offended more traditionally minded rock fans.

In addition to the resistance we encountered on tour, we also experienced a related but slightly different resistance within the music industry. When the Blake Babies became a successful underground band, artist and repertoire (A&R) men began to express interest (only men ever expressed interest; women entering the A&R community seem largely to be a phenomenon of the 1990s). All of the A&R men presented a similar vision for the band: present Hatfield as the sex symbol focal point and supplement the sound with additional (male) musicians. The idea of a sexually integrated band of equals remained far too radical: either a woman fronted a band with a group of competent male musicians backing her up, such as the Pretenders, or else all of the instrumentalists were women, exploiting the all-girl band marketing niche like the Bangles. Our unwillingness to trade on either the novelty or sex appeal of the female rock instrumentalist presented a real challenge to the 1980s mainstream music business.

While the Boston bands mainly played more ethereal or melodic music, elsewhere in the country several groundbreaking female-fronted bands took a much more aggressive, confrontational stance. Two bands that actively challenged gender stereotypes by seeking to out-macho male bands in their own unique ways were Babes in Toyland from Minneapolis and L7 from Los Angeles. Babes in Toyland, featuring the ironic, shrill-voiced singer-guitarist Kat Bjelland, dressed like Sunday schoolers complete with patent leather shoes and baby doll dresses. Their hyperdynamic brand of heavy rock expressed an energy that transcended sexuality. Bjelland seemed to almost taunt her audience, daring them to reduce her to an object of desire. L7 fit more closely into the long-established bad-girl mold, intimidating male audiences with their unkempt, apathetic image. They constantly pushed the envelope of what was acceptable for women rockers: at one legendary Reading festival appearance a member of the band removed a bloody tampon and threw it into the crowd. While L7's antics tended to overshadow its music, it became one of the best all-female hard rock bands ever assembled.

Within this generation of aggressive, female-fronted bands, Hole achieved the greatest cultural and commercial impact. Singer-guitarist Courtney Love truly lived up to Paul Westerberg's (the Replacements) assessment of pretty girls "playing makeup / wearing guitar" in his 1986 song "Left of the Dial." A founding member of Babes in Toyland and former member of Faith No More, Love appropriated the baby doll dresses and heavy, sloppily applied makeup. She frequently stood on stage, microphone in hand and foot on monitor, and simply let her Fender guitar dangle around her neck. She truly embodied the empowerment that came with playing the electric guitar: "I strap on that motherfucking guitar and you cannot fuck with me. That's my feeling."[11] Love depended heavily upon her male lead guitar foil Eric Erlandson, but the rest of her band remained exclusively female throughout several lineup changes. A skilled media manipulator and unapologetic opportunist, Love managed to be in the right place at the right time when the grunge explosion hit in 1991. Hole's second album, *Live through This,* went platinum, and she and husband Kurt Cobain of Nirvana became the most visible and outspoken figures of the grunge movement.

In the early 1990s grunge completely revitalized primal hard rock as a marketable commodity. Unlike other styles of commercial music, many grunge musicians espoused sociopolitical agendas, often rooted in tolerance and inclusiveness. Kurt Cobain used his considerable media voice to promote obscure bands, many of which promoted female guitarists, such as the Raincoats and the Japanese pop-punk trio Shonen Knife. He helped to bring many obscure bands from his native Pacific Northwest to international attention, including mixed-gender bands such as Beat Happening and the Fastbacks. Around the time the first grunge bands leaped from obscurity into the mainstream, another movement centered in the Pacific Northwest (and Washington, D.C.) began to receive media attention. The Riot Girl (or Riot Grrrl) movement remains the only organized, systematic attempt by women to infiltrate the male-dominated world of rock.

The Riot Girl movement began in the early nineties as a loosely organized national network of young women with a feminist agenda who rejected the tactics and goals of the previous generation's feminism. They adopted the attitude and rhetoric of the punks and communicated through the tried-and-true punk tradition of the fanzine. The main target of the riot girls was the increasingly macho hardcore punk scene. The Riot Girl manifesto stated, "Death to all fucker punk boys who refuse to acknowledge the girl punk revolution."[12] They encouraged girls to join or form bands, teaching each other guitar chords and drum beats as if it were paramilitary training to overthrow the government. In the many Riot Girl or Riot Girl–inspired bands of the early nineties, men (such as Billy Karren of Bikini Kill) played supporting roles. Women ran the show, even to the extent of holding performances exclusively for women.

The irony of most Riot Girl bands is that while they espoused their message through their lyrics, their music was inspired by the macho styles of sixties garage rock and seventies punk. The aggressive nature of the music helped to get their strong message across but failed to distinguish itself as a uniquely feminine style of rock. Many Riot Girl bands shunned technical proficiency because they wished to keep their music free from cliché. But unlike 1970s bands that embraced their incompetence such as the Slits and the Raincoats, the Riot Girl bands' music tended to be derivative rather than instinctive. Even the most popular and "revolutionary" bands of the movement failed to deliver timeless albums.

The enormous and often misleading press coverage of the Riot Girl movement created an international phenomenon but effectively defanged the original message. Media blitz aside, the Pacific Northwest, specifically Olympia, Washington, had been a thriving underground pop scene for years. Inspired by original Riot Girl bands Bikini Kill and Bratmobile, guitarists Corin Tucker and Carrie Brownstein formed their own bands, Heavens to Betsy and Excuse 17. Eventually they united to form Sleater-Kinney, which became more musically and lyrically sophisticated—a grown-up version of the primal punk rock of the Riot Girl bands. Sleater-Kinney went on to achieve huge critical and modest commercial success, paving the way for similar West Coast bands such as the "queer core" of the all-lesbian punk band Team Dresch and the introspective indie rock of the Spinnanes. With aggressively independent record labels such as K and Kill Rock Stars actively supporting this movement, the Pacific Northwest continued to produce interesting mixed-gender records long after Seattle ceased to be considered the center of the rock universe.

By the early 1990s the Boston music scene finally produced a series of commercially successful female acts. The Breeders evolved from the relatively successful Pixies and Throwing Muses, with Kim Deal fronting and playing an aggressive amplified Martin acoustic and Tanya Donelly playing lead electric. After the first album Donelly left the Breeders to form the hugely popular Belly, and Deal returned to her hometown of Dayton, Ohio, and enlisted her twin sister Kelley, a beginner guitarist, to fill the role of lead guitarist. Deal acknowledged that "there's some really good guitarists around Dayton, they play so fast, and they know all these blues-jazz riffs. But it's weird—no matter how fast they repeat them, it still sounds so boring and soulless. At least Kelley's an original."[13] Despite the fact that Kelley could barely form chords, she recorded on the Breeders' massive 1993 album *Last Splash* and provided the catchy lead guitar riff that propelled the hit single "Cannonball," proving once and for all that technical proficiency is not a prerequisite for making successful records.

The American rock and pop underground continued to be a healthy breeding ground for female musicians throughout the 1990s. All-female or mixed-gender bands continued to emerge throughout the decade, such as Veruca Salt, Cat Power, Helium, and the Donnas, who revived the teenage bad-girl image in a new, postpunk context. Chicago's Liz Phair produced one of the most articulate neofeminist statements in her 1993 album *Exile in Guyville,* which was based on the ultramacho 1972 Rolling Stones album *Exile on Main Street.* In a move unthinkable twenty years before, Phair addressed her own sexuality with discomforting candor. She wrote all of the songs for the album alone and then assembled an all-male backing band (featuring producer Brad Wood) to record them. In a novel approach, she recorded the electric guitar parts first (with a Fender Mustang) and then commissioned her band to add the backing tracks. The resulting recordings express a rare intimacy within the rock context, demonstrating that women's influences can help to revitalize the style.

With rare exceptions such as Riot Girl band Huggy Bear and the occasional neopsychedelic "shoegazer" bands such as Lush and My Bloody Valentine, female electric guitarists had been absent from the English rock scene since the punk movement fizzled out in the late 1970s. In the mid 1990s a powerfully feminist yet indisputably modern singer-guitarist burst onto the British scene. Polly Jean Harvey led her trio, PJ Harvey, with an unconventional guitar style that provided the emotional backdrop for her stark, affecting songs. Although she denied possessing any real ability as a guitarist, her performances spoke for themselves. After two albums that effectively took the punk-influenced Riot Girl style to a new level of accomplishment, Harvey left her band and returned as a solo performer. She helped to kick start a brief women's rock revolution in the United Kingdom, with bands such as Elastica and Echobelly.

For a brief period during the middle of the decade it appeared that women would completely take over the mainstream as well, with the phenomenally successful Lilith Fair summer tours that exclusively featured female talent and the runaway international success of singer-songwriters such as Alanis Morrissette, Sheryl Crow, and Jewel. But after three years the Lilith Fair tour went into a hiatus and the trend quickly ended, replaced in the mainstream by the unprecedented misogyny of rap-rock groups such as Limp Bizkit and Kid Rock.

Despite the relative absence of women in the current mainstream music scene, certain trends offer a beacon of hope. The Fender custom shop has begun designing electric guitars specifically for women, with thinner necks and slimmer bodies. Bonnie Raitt and Courtney Love have both lent their names to recent designs. While the trend of child prodigy blues guitarists—such as Shannon Curfman and Susan Tedeschi—suggests that the blues has been reduced to mere nostalgia, the presence of women on the scene proves that Bonnie Raitt was not merely a fluke. And most interestingly, as the heavy metal scene has shaken its transgender experimentation and has become more overtly macho than ever, a teenage all-female metal band has found success. Kitty plays extremely aggressive thrash metal pioneered by macho bands like Type O Negative and Korn, but it seems to be accepted by headbangers despite the members' gender.

One compelling piece of evidence that women playing the electric guitar have become fully ingrained within the international cultural landscape is a recent Sony Play Station game called Um Jammer Lammy, created by Masaya Matsurra and released in 1999. Um Jammer Lammy features Lammy, the guitar player in an all-female band, whose mission is to play increasingly intricate lead guitar licks to stymie a series of bad guys on her way to an important gig. While the game itself is intentionally kitsch and slightly ridiculous, the guitar clearly becomes the instrument of physical empowerment for Lammy. With only a small creative stretch, one can perceive the game as a metaphor for the obstacles women rockers (or women in general) face in the real world.

Despite this evidence that women will continue to break down barriers in the rock community in the future, the fact remains that the electric guitar is still largely the domain of men. Few recent female role models have emerged, and rock appears to be more macho than ever. A recent issue of *Guitar Player* asked where all the women guitar players had gone: the Riot Girl movement had burned out and many of the most promising bands had folded. Few in the present day would suggest that guitars look strange on girls. We are now used to seeing women playing electric guitar, and the novelty has been, thankfully, stripped away. While the conservative, cyclical nature of the music business has created a lull in the trend, more innovative women guitarists are likely to follow as the musical landscape changes.

I found that it is nearly impossible to locate willing interview subjects on this topic, because all female electric guitarists seem to have one thing in common: they hate being regarded as *female* electric guitarists. In true feminist fashion, they prefer to be regarded as guitarists who happen to be women. Perhaps this points the way to the real goal of women electric guitarists: to break down the gender barrier to the point where people stop paying attention to the sex of the player and only notice the quality and originality of their performance. As with many traditionally macho institutions, this goal is still a long way off. Since the electric guitar has become so fully ingrained in our culture, the likely scenario is that women electric guitarists will continue to reflect the constantly changing role of women in our society in general.

So what about the elusive great female innovator mentioned by Juliana Hatfield? Is she necessary to affirm the significance of women in their contributions to the development of the instrument? It seems all but inevitable that eventually a woman will come along with such sweeping influence that she will change the way the instrument is played. The fact that she has not come along yet may be nothing more than a fluke, and in view of the incredibly small number of real male innovators, it's not even all that surprising. But considering all of the interesting, brave, proficient, and subtly influential women electric guitarists over the course of the instrument's development, her emergence becomes unnecessary.

Sources

Arnold, Gina. "Women with Guitars: Why Are There So Few of Them?" *Request*, October 1992, pp. 34–36.

Guitar Player magazine.

Most of the interviews were carried out during the 2001 Blake Babies reunion tour.

Notes

1. Patricia Kennealy-Morrison, "Rock around the Cock," in *Rock She Wrote*, ed. Evelyn McDonnell and Ann Powers (New York: Delta, 1995), p. 363.

2. Amy Raphael, *Never Mind the Bollocks: Women Rewrite Rock* (London: Virago, 1995), p. 51.

3. Simon Reynolds and Joy Press, *The Sex Revolts: Gender, Rebellion, and Rock'n'Roll* (Cambridge: Harvard University Press, 1995), p. 224.

4. Legs McNeil and Gillian McCain, eds., *Please Kill Me: An Oral History of Punk* (New York: Penguin, 1997), p. 231; journalist Lucy Toothpaste quoted in Jon Savage, *England's Dreaming: Anarchy, Sex Pistols, Punk Rock, and Beyond* (New York: St. Martin's, 1992), p. 418.

5. Raphael, *Never Mind the Bollocks*, p. xvii.

6. Ibid., p. 103.

7. Ibid., p. 96.

8. Ibid., p. 81; *Behind the Music*, VH1.

9. "Storming the Boys Club," *Guitar Player*, July 1989, p. 96.

10. Jas Obrecht, "Lita Ford, Return of a Runaway," *Guitar Player*, August 1983, p. 38.

11. Raphael, *Never Mind the Bollocks*, p. 31.

12. Reynolds and Press, *Sex Revolts*, p. 323.

13. Jason Fine, "The Breeders, Beginner's Pluck," *Guitar Player*, November 1993, pp. 14–15.

Further Reading

O'Brian, Lucy. *She Bop: The Definitive History of Women in Rock, Pop, and Soul*. London: Penguin, 1995.

Whitely, Sheila, ed. *Sexing the Groove: Popular Music and Gender*. London: Routledge, 1997.

conclusion

the electric guitar at the millennium

In the 1980s the two extremes of the guitar-playing fraternity were heavy metal and (what we now call) alternative.[1] United only in their disdain for middle-of-the-road popsters and antiquated, classic sixties rock, they stood for completely opposed and incompatible styles of electric guitar music. Only the befuddled parent was unable to distinguish between the two: the outrageous glam rock or "hair bands" with their spandex outfits, makeup, and smoke bombs; and the self-absorbed players with a political agenda and dirty jeans. Corporate sellouts with several platinum records to their credit contrasted with "shoe-gazers" who had only torn T-shirts and a few homemade cassettes to show for their years in music.

This division was revealed in their choice of instruments: checkered, custom-made Les Pauls and brightly colored Flying Vs versus ancient beat-up Silvertones and Mustangs. "Big, fancy guitars" had become one of the emblems of lite metal and hard rock, along with long, bleached blond hair and tight trousers.[2] These highly finished guitars, with numerous layers of metallic paint hand buffed to a brilliant shine, provided an appropriate symbol for the financial rewards now available to hard rock and metal guitarists. These acts were responsible for selling millions of records. Most of the big names had contracts with the majors—the multinational record companies that dominated the music business—and some of them had been co-opted by other large business concerns, like beer companies, who were looking to enhance their sales and their corporate image by sponsoring rock concerts. It was hardly surprising that these corporate sponsors actually took over the shape of the guitar for their self-promotion: Dean produced a Budweiser guitar and Hamer custom built a Miller guitar. Each of these instruments came with the name, colors, and logo of its sponsors. The Hamer guitar was also shaped like the Miller beer bottle label.

Even within the hard rock genre important differences were symbolized by guitars. Jeff Johnson, guitarist with Max Panic, made this distinction when questioned by Harris Berger: classic hard rock bands produced rich distortion and expressive sound with top-quality equipment (Fender Strats, Gibson Les Pauls, and Marshall amplification), but the despised glam bands used inferior Hamer or Jackson guitars cut into outrageous shapes, finished with bright colors, and sporting low actions. They could play fast, but the tone was weak and thin, and there was nothing of the excitement or power of the hard rock classics.[3]

The alternative movement of the 1980s was strongly influenced by the punk rockers of the previous decade and their highly publicized stance against commercialism in rock in general and the stars of the majors in particular. Many alternative players rejected shiny, expensive guitars as part of their reaction against the bloated materialism of rock 'n' roll. Any old guitar would do for punk bands like the Sex Pistols, the Clash, and the Ramones: secondhand was good, stolen was okay too. The demise of virtuosity could be seen in their poorly maintained instruments. The punk players went to some lengths to present an appropriately disheveled look for the guitars they used on stage. Joe Strummer of the Clash was somewhat taken aback to find out that he was not the first musician to write political slogans on his guitar—folksinger Woody Guthrie had beaten him to it by about forty years. "This machine kills fascists" was as appropriate in the 1970s as it had been in the 1930s, and it was followed in the 1990s by slogans that addressed current problems, such as "Arm the Homeless."[4]

A bruised and well-worn guitar helped underline the authenticity of rock bands and made a statement equally as powerful as the custom-shop model with a signature endorsement. Stevic Ray Vaughan's Stratocaster was worn down to the wood, emphasizing his bluesman credentials as a hardworking player constantly on the road. Much the same could be said of Bruce Springsteen's or Keith Richards' workmanlike Telecasters.

The emergence of an alternative to mainstream rock also brought forward some alternatives to the guitar hero and the primacy of guitar virtuosity. In some rock camps in the 1980s virtuosity reached its highest peak of expression and adulation. Hard rock and heavy metal had embraced speedy playing, and guitarists coined yet another term to describe the searing strings of notes coming from virtuoso players like Eddie Van Halen: *shred*. The opposing view was that high-speed guitar histrionics were invalid and that slower did not mean inferior; rather, it gave more room for expression and opened the doors for more creative use of tone (and besides, it annoyed some people in the audience).

Alternative players were also ambivalent about the guitar solo, which was enshrined as a central part of the hard rock and heavy metal performance. But flashy solos were for exhibitionists, and the obligatory solo in the middle of a song stood for the showmanship that many alternative players were rejecting. In place of the blazing solo with long lead lines and runs up and down the frets was a textured, rhythmic use of the electric guitar and its various electronic effects that could be massaged into a signature sound. Players like jazz- and reggae-inspired Andy Summers of the Police or the Edge of U2 used timbral and rhythmic devices to craft distinctive solo sounds. Alternative rock in the 1990s did not do away with the solo altogether but provided an alternative that was built up out of layers of different guitar sounds to create a mood rather than a dramatic sequence in the middle of the song. This approach often relied as much on technology—combinations of effects and extensive engineering in the recording studio—as it did on virtuosity. Because of their low-key stage presence and lack of guitar flourishes, players like the Edge and Johnny Marr of the Smiths were sometimes referred to as guitar antiheroes.

While the virtuosos of heavy metal were constantly innovating and looking for new sounds, the rank-and-file metal and hard rock guitarists kept to the status quo: stripped-down power chords, heavy distortion, involved solos, and extreme volume. Guitarists in alternative rock were more open to innovation, especially when it came in the form of a new tone or a new effects pedal.

There were many more guitar effects available in the 1980s as signal processing became a booming niche of the market for instruments and amplifiers. One of the most popular was chorus: this takes the signal and duplicates it with another that is slightly out of tune. It gives the guitar a thicker, larger sound. Introduced by Roland in the 1970s, it gained many adherents in alternative music, where players like Summers and Marr used it extensively to build their guitar sound. Guitar effects had initially been applied one at a time; now they were being strung together to form complex manipulations of the signal. The sequential combination of effects—such as flanger, chorus, delay, and reverb—could be the secret of a new guitar sound. Maintaining the equipment and sequencing the effects called for a degree of technological skill that had been unknown two decades previous. There was hardly any area of guitar-driven music that did not involve any manipulation of tone. Even the punk purists—the players who tried to embody the values and low fidelity of early rock 'n' roll—could not resist effects like chorus and reverb.

grunge

In theory the alternative music community embraced change and diversity, whereas heavy metal adherents wanted to keep to the same sound that Black Sabbath had created years before. The practice was a lot less clear-cut. There was a very wide spectrum of music under the heavy metal umbrella in the 1980s: the ultimate lite metal hair band Poison released "Look What the Cat Dragged In" in 1986, and at the same time Guns N' Roses put out its first album and Metallica issued its influential "Master of Puppets." These recordings were dramatically different in both sound and intent. The alternate scene was moving in so many different directions during that period that categorization would be pointless.

The golden era of heavy metal came to an abrupt end in 1991, when Nirvana released its blockbuster *Nevermind* album and the worlds of pop, alternative, and heavy metal experienced seismic changes and some tectonic shifts. On the face of it a new era in rock music had begun, and it was all over for hair bands, heavy metal, and the singing, dancing pop acts with singles in the Top 40. The new genre soon had an identifying label: *grunge*. Initially the grunge sound had few supporters in record companies or guitar manufacturers; the pioneer bands were making inaccessible, noncommercial music to a very small audience. But the phenomenal sales of *Nevermind* and the follow-up albums of several other bands, incorporating the Seattle Sound, were to change all that. Within a very short time heavy metal was deemed to be passé by the hundreds of bands with the new Seattle Sound and impressive recording contracts with the majors.

Despite the fact that several strains of metal, especially underground black and death metal, proliferated in the 1990s, some commentators announced the end of heavy metal. In the view of Martin Popoff 1991 was "the year metal died."[5] No more would bands like Poison and Def Leppard achieve double platinum sales. Many lite metal acts disappeared without a trace (only to return a decade or so later as nostalgia acts). The hair was out, the outfits were ridiculed, and the sound was obsolete. In its place was the grunge sound: ragged, howling, and totally distorted. It was still a guitar sound, but it was a dense muggy sound, based on riffs instead of solos and often overlaid with bleak, cynical lyrics.

Grunge reestablished the electric guitar as cool and encouraged another wave of amateurs to pick up instruments and emulate a new type of guitar hero. But the look and role of the electric guitar was changed. Its position as status symbol and talisman of the rock 'n' roll dream was reduced. Now it was more a functional tool, as dirty and worn as the clothes the Seattle bands wore on stage. Personalizing the instrument became important, but instead of signature models and shiny, iridescent paint jobs there were slogans, stickers, and graffiti. The grunge players made their guitars look like their battered skateboards.

Kurt Cobain of Nirvana was now the most copied guitar player in America. His favorite instruments were the cheap guitars that he found in pawnshops—an appropriate choice for someone who regularly destroyed his equipment on stage. When interviewed in *Guitar World* in February 1992, he said this about his guitars: "they're cheap, inefficient and sound like crap." This represented a new aesthetic for players and was probably the last thing that the guitar manufacturers wanted to hear. Cobain was in single-minded pursuit of tone, and he found desirable tones in older technologies. Initially the attraction was their cheapness and obscurity, but as grunge became a big business, old equipment rapidly became highly prized and ruinously expensive. Cobain liked to browse secondhand stores for his clothes and his guitars. He also searched out old effects pedals dating back to the 1970s, such as the Electro-Harmonix Big Muff distortion pedal, many of which were no longer available.

None of this was good news for the guitar manufacturing industry. The heavy metal guitar was now as discredited as the bands that played them, and suddenly the brightly painted instruments with pointed Samurai headstocks and Floyd Rose tremolos were no longer selling. New fashions in guitars challenged the supremacy of the Les Pauls and the Stratocasters as well as the Yamahas and the Jacksons made popular in the 1980s. The stars of grunge were like the heavy metal players in their admiration for antique guitars, but instead of Flying Vs and Explorers they were acquiring old Fender Jaguars and Mustangs. Introduced in 1962 and 1964 respectively, these two Fender products had never achieved the success of the Strat or the Telecaster. In the 1970s the Jaguar joined the Jazzmaster at the top of Fender's line, but it was not a best-seller. Two decades later the secondhand price of these models was much less than the classic Stratocaster or Telecaster, and this made them attractive to penniless musicians. The growing popularity of these models during the rise of grunge forced Fender to revive them. In 1999 U.S.-made Jaguars and Jazzmasters appeared on the American market for the first time in twenty years.

Despite his previous comments about crap guitars, Cobain was soon working with the Fender custom shop to create his own signature guitar, the Jag-Stang, a cross between the Jaguar and the Mustang. Instead of a depression in sales, guitar manufacturers and dealers found themselves swept along by a wave of "retro fever" for antique guitars, both originals and reproductions. Guitarists of the twenty-first century maintained the reverence for the classic electric guitars of the 1950s. In a high-tech business constantly striving to come up with something new, the most desirable tool was still the fifty-year-old guitar.

Grunge turned out to be the best thing to happen to guitar makers since Kiss made it fun to rock 'n' roll. In the long history of rock music, no new guitar sound was as extensively and expertly marketed as this. Nirvana's sound became the blueprint for another ten years of me-too groups, all using equipment and effects marketed as grunge. This applied to guitars, amplifiers, and effects. Everything was for sale. If you could not locate a precious Big Muff distortion pedal, you could buy a modern version or an imitation that promised the same sound.

Despite the perceptions of newness and change, grunge actually had a fairly long musical pedigree. Most of the kids who picked up electric guitars in the 1980s (and went on to become the guitar players in the Seattle bands) were far removed from the early influences of rock 'n' roll: Buddy Holly, Duane Eddy, and maybe even Pete Townshend had little meaning to them. They grew up on Kiss, and when they thought about classic rock 'n' roll it was bands like Black Sabbath that came to mind. In fact Black Sabbath provided the links between grunge, punk, and heavy metal. Joey Ramone spoke for a generation of guitarists when he acknowledged, "We grew up on Black Sabbath."[6] Black Sabbath's low, sludgy riffs could be heard in the droning, growling tone of grunge, which had elements in common with both heavy metal and alternative music. When Kurt Cobain wrote out a promotional flyer for his new band, he defined Nirvana's music as "heavy rock with punk overtones," and in his band biography for its first record label, he said it sounded like "Black Sabbath playing the Knack, Black Flag, Led Zeppelin and the Stooges."[7]

Loudness and distortion were probably the only things that these bands had in common, and these two features still defined the sound of electric guitars in the 1990s. Rock music ended this decade the same way it started it—drenched in distortion. Heavy distortion was so commonplace in rock music at this time that some young players attending their first guitar lessons did not recognize the clean unaltered tones of vintage Stratocasters—all they had ever heard was fuzz and hum! Loudness was important to everybody, and the power of the electric guitar and amplifier was still the main reason to play it. Both hair band and grunge devotee agreed that the music had to be loud enough to physically affect the listener. Loud guitar music was also understood to have a therapeutic, transcendental effect on the teenage listener regardless of the allegiances of the players.

Rather than ending heavy metal, the grunge bands reinvigorated it and gave it new clothes and meaning. Bands in the wake of Nirvana, like Alice in Chains or Soundgarden, incorporated enough heavy metal in their sound to make it respectable again. But there was a new dress code for rock guitarists and a new tone to reproduce. Guitarists who wanted to achieve the fashionable low throbbing guitar tone had two options. One way was to tune down the guitar strings. The drop D tuning was most popular among Seattle bands, and this was the most copied. The low E string was tuned down one step to D. One of the advantages of drop D tuning is that it makes it very easy to form chords by just placing a finger across the strings, as in the barre of the barre chord. This makes it attractive to an unschooled player. Another common approach was to tune down all the strings a whole step; E going to D, A to G, and so on. This gives a lower overall pitch range, and the player does not have to learn new chords and scale shapes. Players also used heavier strings to reduce string flap and thicken up the tone. Grunge did not invent drop tuning; it had a history all the way back to early blues players, but now it was widely used in guitar rock.

Another way to achieve the trendy, low-down guttural tone was to use a seven-string guitar, the seventh string being a low B. Thus low to high the guitar strings were tuned B, E, A, D, G, B, E. The extra string made it easier to get the sound at the heart of the ominous, pounding riffs of 1990s rock. Seven-string guitars were not new; jazz guitarist George Van Eps played an Epiphone in the 1930s and Steve Vai used an Ibanez in the 1980s. What was new was an entirely different product for the mass market of amateur players. Naturally the Japanese manufacturers were the leaders in marketing seven-string models, and they obtained the celebrity endorsements of a new generation of guitar heroes. By the end of the 1990s most of the leading manufacturers offered seven-string guitars, ranging from cheap entry-level instruments to top-of-the-line models with celebrity signatures on them.

relics and replicas

The emergence of a new genre of rock music with a different sound provided opportunities for the guitar and amplifier manufacturers, and not the least of these was more guitar-playing celebrities to help market their products. In the early days of the solidbody electric guitar, there were only a few celebrity endorsements, but fifty years later there were scores of special guitars, custom designed to fit the playing style and reflect the image of star players drawn from the entire spectrum of popular music. This trend moved in tandem with the resurgence of custom shops and the status of hand-building quality instruments—a significant trend in the manufacturing industry in the 1980s and 1990s.

The custom shop represents the advantages of small businesses in American manufacturing: high-quality work, great flexibility, and personal attention to customers. Companies that had benefited from heavy metal patronage, like Hamer and Dean, were able to increase the scale of their operations in the 1990s, moving into larger production runs and marketing their products to a wider customer base. At the same time the larger, established concerns like Fender and Gibson went in the other direction—setting up custom shops to produce a limited number of instruments at the top of the line. This was part of their strategy to revive the reputation of American manufacturers after the bad press of the 1970s. The large, impersonal guitar manufacturer, often part of an even larger consolidated concern, was not viewed favorably by the people who bought guitars. The custom shop, however, represented a caring and sympathetic attitude toward musicians.

The booming market for antique guitars was proof enough that players were not satisfied with the quality of mass-produced guitars. Custom shops covered both bases by offering hand-built instruments and high-quality copies of the classic guitars. Fender's custom shop found a very strong demand for "replicas" and "relics" (a relic was a replica that had been weathered to look used) of 1950s Telecasters and Strats. The concept of producing New Old Stock (NOS) turned out to be more successful than anyone at Fender had imagined, and soon other manufacturers were reviving their classic designs.

The growing prestige of the custom-shop operation gave some small businesses the avenue for rapid growth. The one important new electric guitar manufacturer in the United States in the 1990s came from this background. Paul Reed Smith was a luthier with a workshop in Annapolis, Maryland. He started by building a guitar as a school project, and five years later he was making custom instruments. When he decided to build an electric model, he looked to the classic Gibson Les Paul, just as numerous other designers had done when they moved into electrics. Smith set out to build an electric solidbody instrument along the same lines as his painstakingly constructed acoustics. He sought out the best materials—aged Honduras mahogany and Brazilian rosewood—and contoured the shaped body. He also simplified the wiring and obtained patents for his bridge and tremolo design.

Paul Reed Smith guitars gained their reputation from celebrity endorsements, in this case Carlos Santana, and soon Smith looked to increase production from one to eight guitars a month![8] This was way back in 1982, when the organization only had a handful of employees. In 1986 the PRS company was formed, and by

Ted McCarty and Paul Reed Smith: two generations of American inventors. McCarty is caressing the PRS guitar named after him. Courtesy of PRS Guitars.

the 1990s PRS guitars were produced in much greater numbers but still kept an enviable reputation for high quality. The company was now making a line of solidbody electrics, including double- and single-cutaway models, and found customers across the spectrum of popular music, including many grunge and heavy metal players. One of Smith's idols was Ted McCarty, and the Lemelson Center's "Electrified, Amplified, and Deified" program brought the two together in a symbolic union of the past and the future of American electric guitars.

The Ventures having fun performing at the "Electrified, Amplified, and Deified" symposium at the Smithsonian. Another indication of musicians' creativity in getting new sounds out of the electric guitar. Photo by Eric Long, Smithsonian Institution, no. PCD2284-001.

As PRS became a major manufacturer, it had to consider moving some production out of the United States. In the 1980s rising costs led many manufacturers to begin offshore production, first in Japan, then (as the yen rose in value) in Korea, and finally in China. Mexico was another popular location to build cheap guitars for the American market. By the 1990s Korea was the major producer of electric guitars, and the world's largest manufacturer was the Korean company Samick. With so many guitars fabricated outside the United States, U.S. manufacturers began to stress the American origins of their products. Fender introduced the American Standard line of U.S.-made guitars, which proved to be extremely popular. This line was superseded by the New American series at the turn of the century, when any guitar labeled "American Made" came with expectations of higher quality.

new faces, new sounds

The solidbody electric guitar entered the twenty-first century with all of its power and prestige in popular music intact. A new generation of guitar heroes have taken over the responsibility of maintaining the musical traditions and preeminent position of the instrument. In an article explaining the appeal of Jimi Hendrix, the musician and cultural critic Frank Zappa pointed out that the females in the audience were taken by his sensuality and beauty, and the males "settle for vicarious participation, and/or buy a Fender Stratocaster, an Arbiter Fuzz Face, a Vox Wah Wah pedal, and four Marshall amplifiers."[9] Nowadays an observer at a rock concert might notice that these gender roles can be reversed. In the twenty-first century female players provide many of the celebrity endorsements required to market electric guitars. Courtney Love has enough star power to justify her own guitar. She codesigned the Fender Venus models, which were inspired by Rickenbacker and Mercury guitars. Jennifer Batten has her own signature guitar made by Washburn. Female guitar players can now be seen on television advertising (for Schlotzsky's Deli), and Joan Jett's music is now promoting information services instead of sex, drugs, and rock 'n' roll.

The spectacular, and unexpected, rise of several teenage guitar wunderkinder in the 1990s included several female players, such as Shannon Curfman and Susan Tedeschi, who now appear as established guitar-playing stars. That many of these new faces learned their skills playing the blues is testament to the popularity and staying power of this form of guitar music. While some of the founding fathers have, regrettably, passed on, their heritage is safe in the hands of a new generation of young players—Kenny Wayne Shepherd, Jonny Lang, Derek Trucks, and Keb' Mo'—who play all around the world and promising young players like Debbie Davis and Deborah Coleman, who work on the club circuit.

Judging by the concert calendar, sixties-style guitar-centered rock has lost none of its mass appeal. The new bands of the 1990s might have been light years distant from the rock 'n' roll dinosaurs on the stadium circuit, but they did not depart from the long-established format of several guitarists and one drummer and they still based their sound on the same electric guitars. Even the new forms of popular music that once threatened to provide an alternative to four guys playing guitars have been colonized by this ubiquitous instrument. The mesmerizing riffs that powered the first rap records, such as "Rapper's Delight" by the Sugarhill Gang, came from electric guitars. Rap might not have been as guitar-centric as heavy metal or alternative, but it still depended on that sound, especially for the rhythm. Many of its practitioners were DJs, and the ready availability of records made it cheaper to use recordings of guitar sounds than to pay a musician to produce them. Early rappers liked to lift samples from heavy metal guitarists like Van Halen to use in their songs. Tone-Loc used a piece of Van Halen's "Jamie's Cryin'" on his hip-hop song "Wild Thing." Rick Rubin, the producer behind the Def Jam record label, took samples from bands like AC/DC, Led Zeppelin, and Slayer to use in hip-hop records.[10]

When rap pioneers Run/DMC joined hard rockers Aerosmith on the celebrated *Walk This Way* video in 1986, they were doing more than celebrating the collegiality of rock 'n' roll; they were pointing the way to the future. In the twenty-first century there are rappers with guitars and hard rockers who rap. In the late 1990s hip-hop and heavy guitar came together in immensely popular bands like Rage Against the Machine and Limp Bizkit. Rap metal is now one of the most successful strains of the postgrunge rock scene. Several hip-hop artists migrated back to rock at the beginning of the twenty-first century, because they found the instrumentation and the format more versatile and more expressive. An article in the *New York Times* that described this movement was entitled "Hip Hop Generation Grabs a Guitar," underlining the electric guitar's role as the defining symbol of rock culture.[11]

The growing popularity of electronic dance music—with its genres of house, industrial, and techno—presented more of a challenge to the hegemony of the electric guitar in the 1980s and 1990s. After the successful introduction of analog synthesizers in the 1970s, the next generation of digital technology was incorporated into sequencers, digital samplers, and software that allowed a personal computer to control all these elements. This was a cheap, easily mastered, and extraordinary versatile instrument—not an instrument in the conventional sense but an electronic system without boundaries. Taken together, these digital machines could do all an electric guitar could do and much more. Advocates of the new technology believed that they had found a replacement for the ubiquitous electric guitar and all the guitar bands and all their music; the guitar had been "reinvented" in a more personal and efficient form. For the many players and composers in electronic music, the motto No Guitars had real meaning.

This is not to say that there could be nothing in common between guitar players and musicians poring over sequencers and computers. Consider this description of a group of musicians: "Unencumbered by the demands of the critics, fueled by the siege mentality of the true believer, and propelled into complexity by the increasing power of their equipment."[12] This does not refer to heavy metal players in the late 1970s but hardcore dance music creators in the 1990s! Electronic dance music might be assembled in the home studio in a digitally sequenced time frame, but when it moves to the club or the concert hall and becomes a performance, it often incorporates instruments like the keyboard, the turntable, and the old reliable electric guitar.

The computer-controlled electronic music makers of the 1990s promised to put an end to power chords and blues riffs but ended up by enhancing them. The DJs and computer freaks who constructed this most modern of popular music still used the guitar sound in their assemblages of prerecorded sound, albeit in sampled form. The musician of the twenty-first century might not have to learn to operate any other machine than a computer or sequencer, but he or she still leans heavily on the sound of the electric guitar—a sound that is sampled or copied and then added to the mix. One can still hear the power chord and the heavy metal riff in electronic popular music—only it has been reproduced and manipulated in the digital mix. Listening to an album like the Beastie Boys's *Paul's Boutique* is like going down the guitar player's memory lane. One song aptly named "The Sounds of Science" contains guitar samples from the Beatles' *White, Abbey Road,* and *Sgt. Peppers' Lonely Hearts Club Band* albums. On the same track you can also hear samples of funky guitars from Isaac Hayes's *Shaft* soundtrack and from James Brown.

In a book about electronic synthesizers, Paul Theberge made the important point that new technology changed musical practice from production to consumption, that choosing the right piece of sound had become as important as creating it. This is certainly the case with electronic music, but it also applies, to a lesser extent, to guitar players. The rapid diffusion of digital technology has produced a much wider range of choices for the guitar's sound, and many of these choices are modeled on the sounds of the past. The same technologies used in electronic dance music have been applied to effects boxes and amplifiers for guitar players. The digital amps that can now recreate the sound of numerous amplifiers and effects (through modeling algorithms that process the incoming signal and replicate the timbre of classic models) offer an endless variety of tone and effects. "Dialing up a tone" has become a very complex, learned set of practices that determines the sound of the electric guitar.

Charlie McGovern has already made the point that rock 'n' roll is fundamentally a recombinant music, made all the easier by the versatile electric guitar. Recycling the past is common to practically every type of popular music described in this book, but in the twenty-first century it is being done literally—with samples of old recordings or copies of classic guitar amplifier tones—rather than copying or reinterpreting a style on the same instrument. One of the great advantages of the new digital musical order is that the creator (here called the DJ) is not limited by era, playing style, or proficiency on the instrument, but only by the depth of her or his knowledge of old records and historic sounds. In this way not only is the electric guitar sure of survival in the twenty-first century, but also its sonic history is enshrined in the samples and tones that are incorporated into a wide range of popular music. DJs who have reproduced memorable guitar sounds of the past, like surf or power pop, have exposed them to a new audience, assuring that the most influential and versatile instrument of the twentieth century will have a bright future in the twenty-first.

Source

Page, John, vice president of Fender's Custom Shop. Interview by author, 15 October 1997.

Notes

1. The term *alternative* achieved wide currency in the 1990s rather than the 1980s, but it is used here as a shorthand for a very diverse group of guitar players ranging from hardcore punk through jangly Britpop and arty funk. The term was used as a marketing category to differentiate between music produced by the major record companies and that of the independents, which meant that alternative was not heard as much on mainstream commercial radio and did not appear on the large concert circuits.

2. Quote from Dawn Anderson, *Backlash* magazine (Seattle) in the movie *Hype* (1996), directed by Doug Pray.

3. Harris M. Berger, *Metal, Rock, and Jazz: Perception and the Phenomenology of Musical Experience* (Hanover, N.H.: University Press of New England, 1999), p. 52.

4. This is what Tom Morello of Rage Against the Machine has inscribed on his electric guitar, along with some children's stickers and a part of the design of the Soviet flag.

5. Martin Popoff, *20th Century Rock and Roll: Heavy Metal* (Burlington, Ont.: Collectors Guide, 2000), p. 130.

6. *Alternative Rock: From the Pages of Guitar World* (Milwaukee: Hal Leonard, 1999), p. 88. This is a collection of interviews published in this magazine, including Kurt Cobain's 1992 interview.

7. Michael Azerrad, *Come As You Are: The Story of Nirvana* (New York: Doubleday, 1994), p. 108.

8. *Guitar Player*, March 1982, pp. 24–29.

9. Steve Waksman, *Instruments of Desire* (Cambridge: Harvard University Press, 2000), pp. 192–93, quoting Zappa's 1968 *Life* article.

10. Ulf Poschardt, *DJ Culture* (London: Quartet, 1998), p. 269.

11. *New York Times,* 11 August 2002, sec. 2, pp. 1, 28.
12. Peter Shapiro, ed., *Modulations* (New York: Caipirinha, 2000), pp. 134–35.

Further Reading

Redhead, Steve. *The End-of-the-Century Party: Youth and Pop towards 2000.* Manchester: Manchester University Press, 1990.

Theberge, Paul. *Any Sound You Can Imagine: Making Music/Consuming Technology.* Hanover, N.H.: Wesleyan University Press, 1997.

contributors

James P. Kraft is associate professor in the Department of History at the University of Hawaii, Manoa.

Charles McGovern is a curator in the Division of Cultural History, National Museum of American History, Washington, D.C.

André Millard is professor of history and director of American Studies at the University of Alabama, Birmingham.

Susan Schmidt-Horning was recently awarded a Ph.D. in history from Case-Western Reserve University in Cleveland.

John Strohm is enrolled at the Cumberland School of Law, Samford University, Birmingham, and is also working on his third solo album.

song title index

general index